Making Babies

Making Babies:

The Science of Pregnancy

DAVID BAINBRIDGE

HARVARD UNIVERSITY PRESS

Cambridge, Massachusetts

2001

Weidenfeld & Nicolson published an earlier version of this book in the United Kingdom
in 2000 under the title *A Visitor Within: The Science of Pregnancy*.

Library of Congress Cataloging-in-Publication Data

Bainbridge, David
Making babies: the science of pregnancy / David Bainbridge.
p. cm.
Includes bibliographical references and index.
ISBN 0-674-00653-4 (alk. paper)
1. Pregnancy. 2. Pregnancy--Physiological aspects. I. Title.
RG558 .B35 2001
618.2'4--dc21
2001024166

For Michelle, Eleanor and Bump.

A-ha! So it does work!

Contents

○ ○ ○ ○ ○ ○ ○ ○

And I will put enmity between thee and the woman and between thy seed and her seed; it shall bruise thy head, and thou shalt bruise his heel.

Unto the woman he said, I will greatly multiply thy sorrow and thy conception; in sorrow thou shalt bring forth children.

Genesis iii, 15–16

Prologue

○ ○ ○ ○ ○ ○ ○ ○ ○

I got the call at 2.30 in the afternoon. I was setting up a machine to characterise a particular human gene that I was studying, but that would now have to wait. I knew that Michelle was seeing her midwife that afternoon, but all her previous visits had been uneventful, so I had put the visit to the back of my mind. Ever since we had nearly lost the baby when Michelle went into premature labour at twenty weeks, we had both felt a mounting sense of relief as the kicking bump had stayed resolutely put for the next ten weeks. At thirty weeks, I think we both thought that we were home free.

But Michelle was sobbing. All she said was 'I've got pre-eclampsia', and I picked up my car keys and left. I drove home along the Ridgeway wondering what would happen now. A little knowledge is a dangerous thing, and I had a little knowledge about pre-eclampsia. That was the irony of the situation – for the last year I had been working in a lab where almost everyone else was working on pre-eclampsia. So I knew that it is the main cause of maternal death in Britain, killing about ten women a year. I also knew that it leads to the death of five or six hundred babies a year. It attacks mothers' blood vessels, forcing their blood pressure to rise, damaging their kidneys and causing sudden fits – *eclampsia* is Greek for 'flash of lightning'. Pre-eclampsia can also slowly strangle the blood vessels that connect the baby to its mother, gradually starving and asphyxiating it. Yes, I knew all the things that could happen, but I just didn't know how likely they were.

I

I was home in half an hour and then we were on our way to the Radcliffe Hospital in Oxford. I knew the way, as my main lab was on level 3 of the Women's Centre there. I had not yet realised that Michelle was going to spend much of the coming months on level 5. Although I tried to reassure her, we were quiet for much of the journey. We had fallen into the usual trap of complacency. It was as if we had thought that the mere act of having all those scans, all those visits to the doctor, would somehow ward off any problems. I can't help feeling that we had treated the twenty-week ultrasound scan more as a way to get to know our baby than as a way to find out if anything was wrong. Now we were shaken, and everything was uncertain once more.

When we arrived at the hospital, the omens were good. Although Michelle's blood pressure was up, there was no protein in her urine and the baby's heartbeat seemed to be rattling along purposefully. It didn't seem to realise that anything was wrong. Yet the staff at the Radcliffe are careful – some would say obsessive – about dealing with pre-eclampsia, so Michelle was admitted for the night. I did all the usual lone-husband things: I drove home, overfed the cats, packed an entirely inadequate overnight bag for Michelle and drove back to the hospital. Michelle was perched on her bed, fat and forlorn, not knowing whether to be scared or bored. Nothing had changed – her blood pressure remained high, but she still had no symptoms. It was early December, but the overenthusiastic hospital heating system had made her room stifling. Would we get to Christmas?

Pre-eclampsia is not rare – it affects between 5 and 10 per cent of all pregnancies, and perhaps 2 per cent are affected severely. While 'eclampsia' is a vague term for anything that makes pregnant women have fits, 'pre-eclampsia' is a well-defined syndrome. It happens when an unknown substance in the blood of a pregnant woman starts to damage her blood vessels. The vessels react to this damage by constricting, and this makes the woman's blood pressure go up. The mystery substance can also affect the fine blood vessels in the kidneys, so that they start to leak protein into the urine. Blood vessels around the body may also lose protein into the tissues, and this causes swelling of the arms, legs and face. If it happens in the brain, however, the mother can start to have fits and may even fall into a coma and die. This is what is known for certain about this strange disease – somehow pregnancy is attacking the mother's blood vessels.

Pre-eclampsia can be treated, but not cured, by drugs; they simply reduce the symptoms. Drugs like nifedipine and methyl-dopa can help to reduce blood pressure, and valium can also reduce the chances of fits. Unfortunately, we do not know the root cause of the disease, so these treatments are the best we have. Sometimes these drugs are not enough and the mother's condition worsens. When this happens, all that is left is the ultimate cure – to deliver the baby. The baby's presence is the cause of pre-eclampsia, so removing it is the solution to the disease. In most cases, pre-eclampsia starts after thirty weeks of pregnancy, so the baby usually survives if it has to be delivered. Yet some cases can start as early as twenty weeks, when the baby has far less chance of survival.

Michelle came home the next day, stable but not better, and with strict instructions to rest. As Christmas approached, we returned to the hospital again and again. Sometimes her blood pressure was normal. Sometimes it was dangerously high. Always she felt fine, and always the baby's heart pounded away enthusias-tically. It hated the straps of the fetal heart monitor and always tried to kick them off. We were to thank that baby again and again for its bloody-minded resilience. There seemed to be no pattern to Michelle's blood pressure – we would ensure that we reached the hospital in good time so that she was relaxed for her visit, but it never seemed to make any difference. Her blood pressure rose and fell as it liked.

As the weeks went on, things got worse. Her blood pressure gradually rose until she was given nifedipine, and then it rose some more until she was given methyldopa. This made her a depressed insomniac plagued by nightmares, but at least it kept her blood pressure down. But the baby plodded on completely oblivious as Christmas and New Year passed. With January came the first swollen fingers, the sign that Michelle's system was finally giving in and allowing itself to show some symptoms. Next she began to see flickering lights, a sign that the blood vessels in her brain were being damaged. Her stays in hospital became longer and longer as the weeks went on; thirty-four, thirty-five, thirty-six.

Despite decades of study, no one knows what causes pre-eclampsia – how a baby actually damages its mother's blood vessels. Pre-eclampsia affects an intriguing selection of women, however, and this has led to most of our theories about what causes it. It is commoner in women's first pregnancies, especially if they are older when they first become pregnant. In successive pregnancies, the risk of pre-eclampsia gets smaller and smaller, although it never disappears. Because of this, it has been claimed that the disease happens in women whose wombs are somehow 'underdeveloped' when they first become pregnant. Yet a few women get pre-eclampsia in their second pregnancy, but not in their first, so this cannot be the whole answer.

It also seems that women who already have children become more susceptible to the disease again if they start a new family with a new father. It is almost as if their pregnancy slate is wiped clean by the new partner. Whatever causes pre-eclampsia, it almost seems to 'remember' who has fathered the woman's children in the past. Not many parts of the human body are able to 'remember' things like this, but the immune system can, and many people are currently studying the possible role of the mother's immune system in pre-eclampsia. Does the female body remember who has fathered its children in the same way that it remembers whether it has been vaccinated for measles?

Pre-eclampsia remains an enigma, and it remains a killer. Frustratingly, it is also a uniquely human killer, as the disease has not been shown to afflict any other species. This is probably one reason why we have made such slow progress in our attempts to understand it. All we can say for sure is that pre-eclampsia is caused by some failure in the normally smooth process of pregnancy. A baby is entirely dependent on its mother before birth and it continually has to struggle to maintain her cooperation. Pre-eclampsia is a sign that the baby is losing its battle to keep its privileged status as a visitor. If the baby fails it will die, and its mother could die too. Although we often do not realise it until things go wrong, pregnancy can be a very fragile thing.

At thirty-eight weeks, Michelle's obstetricians decided to quit while they were ahead. She was induced, and Eleanor screamed her way into the world at 2 o'clock in the morning of 3 February 1998, wondering what all the fuss was about.

Introduction

○ ○ ○ ○ ○ ○ ○ ○ ○

Pregnancy is a uniquely intimate relationship between two people. All of us luxuriate in this relationship once, and half of us are lucky enough to be able to do it all over again a second time, from the other side, as it were. Never again outside pregnancy can we be so truly intertwined with someone else, no matter how hard we try. In our impersonal, high-technology world, pregnancy remains the one visceral process in which we all take part.

This tiny, red-faced baby that has just landed in our laps and looks as if it is trying to scream itself inside out – what sort of journey has brought it here? How can we make one of these amazing little people, seemingly out of nothing? These are questions that people have asked throughout history: not just the great thinkers among us, but ordinary people like you and me. The miracle of pregnancy – for that is what it is – is one of the few mysteries that almost all of us have pondered. Pregnancy is a very democratic scientific challenge and it can inspire anyone.

This is why I decided to write this book. Ever since I first started to study pregnancy, I have been surprised by the interest that non-scientists have shown in my work. So, I have written this book for people with little or no scientific training who would simply like to know how pregnancy actually works. In my experience, a few major questions crop up in most people's minds when they wonder how we make our children. I like to call these the five 'big questions' of pregnancy, and I have made them the scheme for my book. Each chapter describes our attempts to answer one of these questions – to tell us a little more about the process that makes every one of us.

First of all, people have long wondered why we reproduce in the way we do. Obviously, humans breed by having sex, by making sperm and eggs, and by a woman becoming pregnant, but why? Until the twentieth century, we simply had to accept that this is how people make children, and we often dignified this acceptance by making sex, fertility, male and female into primal forces in our myths of how the universe conducts its affairs. In the last century, however, science has told us the answers to all these questions, and this is the subject of my first chapter. We now know the reasons why there are men and women, why they have to unite to make a baby, which parent contributes more to the baby, and why it is women who get pregnant.

The first time that most couples know they are to become parents is when the little line on the pregnancy test turns blue. After the initial shock has subsided, many then wonder at the changes that are going on inside the mother's body. She is no longer alone – she has a visitor. Although this visitor is tiny, it can completely take over its mother's body almost straight away. In Chapter 2, I will recount how a Renaissance doctor first wondered how the baby manages to secure its own future by stopping its mother menstruating. We will see how biologists had to crack the hormonal code that controls women's menstrual cycles before they could find out how the embryo stops these cycles. The insidious take-over of a woman's body by her baby is the cause of many of the hardships of pregnancy, including breathlessness, anaemia and morning sickness.

Many parents are amazed at how something as intricate as their baby can be put together in the space of nine months, and in Chapter 3 we will see how this is done. Since the ancient Greeks first speculated about how a baby is made from a formless mass of tissue, there have been many theories about what drives embryonic development. In fact, much of the history of biology has been driven by the quest to find out how a child is constructed. Over the centuries, it has gradually become clear how, within the first seven hectic weeks of pregnancy, the embryo changes from a single cell

7

into a recognisable baby with eyes, ears, fingers and toes. During this scientific quest we have found out not only how we are made, but also why we are made this way.

One of the most remarkable things about pregnancy is that it ever succeeds at all. Animals have spent most of their evolution trying to stop themselves being exploited by parasites, and to do this they have developed a formidable array of weapons to destroy would-be invaders. Yet female mammals, including women, have had to turn this defence policy on its head so they can become pregnant. The fetus is a foreign being – half of it comes from its father – but even so, it is not attacked by its mother. The story of how a baby avoids being treated like a parasite or an organ transplant is told in Chapter 4. The coexistence of mother and baby is a triumph of the natural world, but it can have strange unforeseen effects – it can make the mother vulnerable to fatal diseases, and it may be the cause of homosexuality in men.

Finally, birth is the climax of pregnancy – the moment at which the baby must make its bid to survive in the outside world. For much of human history, childbirth was the most dangerous time in a woman's life, and all too often it ended in disaster for the baby as well. Humans drew the short straw as far as birth is concerned – women have been left the evolutionary legacy of babies with enormous heads, and a pelvis only recently adapted for walking upright. Although birth is now much safer in many countries, it remains a fundamental turning point in the life of both mother and baby. The baby must quickly adapt to survive in a completely alien environment – gasping, suckling and clinging its way into life. The mother must turn into a nurturing machine, hell-bent on giving her child the best possible start. A woman's entire life story is designed with the express aim of making a success of pregnancy and birth. Menstruation, menopause, the pain of human childbirth – all these burdens are now thought to exist for the sake of that tiny baby.

Throughout history, human pregnancy has been notoriously difficult to study for many different practical and ethical reasons. Yet the realisation that humans are animals just like any other has made our progress much easier. The world is crammed with millions of different species of animals, each reproducing in its own distinctive way – a veritable treasure-trove of reproductive possibilities. All through this book, I will compare human pregnancy with other animals, especially animals that do things differently. Although studying human pregnancy in isolation can sometimes tell us how things work, a more 'zoological' approach can often help to explain why we make our children in the way we do. Only by discovering what sort of animals we are can we start to explain the idiosyncrasies of how we breed.

Since I started to work on the biology of pregnancy, I have found that it has a unique 'feel' to it – it is where our deepest emotions and driest analytical inquiries collide. Working all day in the lab on some placental gene or other, and then coming home to see my wife's pregnant belly swelling ever larger, made me realise that pregnancy can be one of the hardest things to be objective about. It is never just a science – it is an integral part of our lives that has been woven into our bodies, our history and our mythology. Most of all, pregnancy is a story, and this is why I have written this book as the chronology of a single pregnancy from conception to breastfeeding.

Within the last twenty years, we have finally learnt enough about pregnancy for almost all of its story to be told. Now, for the first time, a coherent story of pregnancy can be told in a form that anyone can understand. This is why I believe this book is important: it is a guide to the 'hows' and 'whys' of pregnancy, at a time when pregnancy is about to become a controversial and emotive issue. Modern technology will soon change pregnancy for ever, possibly even making it redundant. This change will affect all of us, and we need to know how pregnancy makes people if we are to make the right choices about whether to let people change pregnancy.

Yet despite the impending furor about how we can change pregnancy, this is a book about pregnancy 'pure and simple'. Although a great deal has been written in the last few years about surrogacy, designer babies, human cloning and so on, I hope this book will convince you that natural pregnancy is far more interesting than any crude tinkerings that scientists have so far attempted. Nor does this book deal with the politics of childbirth, although many of the scientific findings that I discuss inform our attitudes and treatment of pregnant women and their babies. As far as drama is concerned, natural pregnancy has it all: sibling rivalry, a battle of the sexes, questions of gender identity. We will see that studying pregnancy can help to tell us why people behave the way they do, and why so many of us want to have children. It can even help us to understand who we are. After all, each of us is a little miracle, the product of a million-to-one coincidental meeting of one sperm and one egg that burgeons into a living, breathing person. Although everyday life may make us forget it, this chance encounter is at the root of each one of us.

○ ○ ○ ○ ○ ○ ○ ○ ○

Why do we reproduce
in the way we do?

○ ○ ○ ○ ○ ○ ○ ○ ○

CHAPTER I

Origins

ooooooooo

The race is on. Two hundred million sperm are hurled into foreign territory. After days of waiting, they have been pumped from the chilly, torpid testicle into a warm vaginal haven. Vivified by the cocktail of stimulants poured over them by the prostate gland, they spring to life in the darkness and start their journey upstream, a journey being made within millions of women as you read this page.

If sperm are to reach their goal, they will have to swim a distance 2000 times their own length; the equivalent of you or I swimming 3 kilometres. But swimming is what sperm are for – a sperm is nothing more than a tiny bag of genes and a propulsion system with sufficient stamina to power it to its destination. Although sperm may all look the same, they all carry different genes: different sets of instructions for constructing a baby. In the next few hours, only one sperm can fulfil its potential and reach the egg, and the contents of that sperm will dictate the sex of the baby, as well as many of its characteristics. If the leading sperm falls back and another overtakes it to reach the egg, then the baby that results will be a completely different individual. In an almost inconceivably important few hours, the identity of each of us is decided by the outcome of this tiny spermy swimming race.

And swim they do. The sperm quickly escape from the gelatinous mass that congealed around them as they landed in the vagina and start to swim towards the narrow opening of the cervix. The mucus-filled cervical canal is a major obstacle, and within minutes millions of wiggling sperm are attempting the difficult passage to the other side. Most sperm will fail at this hurdle, but if the time is right, a lucky few will break through into the uterus. No one knows how sperm know which way to go – maybe they swim towards chemicals drifting down from the egg. Possibly they all swim at random, and getting to the cervix is just a matter of luck.

Just when the cervix was making life difficult for the pioneering sperm, help comes from an unexpected quarter. Nerves from the woman's faraway brain start to drive ripples of muscle contraction through the vagina and uterus. These gradually increase until they reach a frenzy and the organs through which the sperm are swimming are wracked with waves of orgasmic contractions. These slow, powerful undulations draw more and more sperm towards and into the cervix, speeding them on their way towards their longed-for assignation with the egg.

The longest leg of the sperm's journey is through the uterus, and as they embark on this stretch, the omens are good. Not only did the cervical mucus allow them through, but the lining of the uterus is heaped into an exuberant mass of ruddy glandular flesh, apparently ready to absorb a new baby into its midst. The ripening uterus is a sign that these sperm may have arrived at exactly the right time – only when an egg is ready to be fertilised is the uterus in this receptive state. Spurred on by their freedom from the cervix, the sperm swim onwards and upwards, ever closer to the end of their journey. Although more and more sperm fall by the wayside, grinding to a halt or thrashing helplessly into the walls of the uterus, a few climb inexorably onwards. Eventually these survivors reach the openings of the two narrow Fallopian tubes that exit from the upper end of the uterus. One of these tubes, either the left or the right, may contain an egg.

Once in the Fallopian tubes, the environment through which the sperm swim changes once more. The walls of the tubes are lined with lawns of tiny hair-like cilia, which beat to and fro, wafting a gentle flow of fluid down towards the uterus. The sperm have to swim against this current, but it is so slow that they make rapid headway. Also, they can now detect chemicals being washed down the Fallopian tube which tell them which way to go. At last, the few sperm that have reached the correct tube are now confronted by the object of their quest, the egg. The egg is still wrapped in the cumulus ('cloud'), a remnant of the cells that fed and nurtured it as it grew in the ovary. A day or so ago, the large sticky cumulus and its valuable eggy cargo were seized by the Fallopian tube and drawn in to start a slow cruise to the uterus. Although large, the cumulus is rather loosely constructed, and the sperm have little trouble swimming through it.

The sperm now reach a more challenging barrier to their progress, which goes by the beautiful name of the zona pellucida ('clear zone'). The zona pellucida is the most unusual of the egg's vestments – a clear glassy sphere with a structure rather like a crystal, and certainly unlike anything else in the human body. The outside of the zona bears a coating that allows the sperm to attach. When the sperm touches this coating, not only does it stick, but also it undergoes a complete transformation. The gentle, sweeping movements of the sperm's tail that smoothly propelled it to the egg are now replaced by frenetic, staccato thrashing movements that thrust it into the zona. Also, the front end of the sperm discharges a potent cocktail of chemicals on to the surface of the zona that starts to attack it. The combination of these chemicals and the sperm's frenzied movements allow it to drill a tiny tunnel through the zona and slip through to the space between the zona and the egg itself. This space is extremely cramped and so even the sperm, tiny as it is, is squashed flat against the egg.

Just at this point, where the victorious sperm is finally able to fuse with the egg, a reversal of fortunes takes place and the egg becomes the active agent in fertilisation, controlling the future of the now passive sperm. The first thing the egg must do is prevent the entry of a second sperm – a situation called polyspermy. Polyspermy would be a disastrous turn of events because human babies are simply not meant to be made from two sperms and one egg. Such embryos are occasionally generated by accident, but they cannot cope with too many genes and so polyspermic embryos are doomed.

Because of its terrible consequences, polyspermy now becomes the egg's temporary obsession, and it has two ways of avoiding it. First, as soon as the sperm starts to fuse with the membrane of the egg, the electrical behaviour of the egg starts to change. Near the sperm's entry point, an electrical current flows across the egg membrane, and this alters the voltage across the membrane. This voltage change propagates out from the sperm to spread over the entire surface of the egg. Within seconds, the chemistry of the egg membrane is irreversibly changed, rendering it resistant to further sperm binding. This is the 'fast block to polyspermy'. The egg's second strategy is the 'slow block' and this takes place when the voltage change of the first block lets calcium atoms leak into the egg from the surrounding fluid. This calcium causes the egg to empty hundreds of special little bags of chemicals into the space between the egg and the zona. These chemicals now alter the molecular structure of the zona to make it resistant to the entry of other sperm.

The egg was desperate to prevent more than one sperm getting in, because the outlook for embryos with too much genetic information is grim. However, just because the egg has prevented polyspermy, the threat of genetic overload is not over, and now the culprit is the egg itself. At the time the egg is fertilised, it contains twice as many genes as it needs to make a normal embryo. Any failure to pare down its genetic content would be just as disastrous as allowing extra sperm in. Because of this, the egg spends its first few post-fertilisation hours packing half of its genes into a tiny parcel called a polar body, which it then ejects.

Now that the egg has resolved these pressing genetic problems, fertilisation proceeds apace. While the egg has been extruding the polar body, it has also been busy dismantling the sperm. The sperm's head is starting to fall apart and its tail has been snapped off and can be seen drifting around inside the egg. The male and female genes now congregate into two separate bodies within the egg, the pronuclei, and these move together and fuse, allowing the paternal and maternal genes to mingle. The egg is no longer just an egg, but is now called a zygote in recognition of its fundamentally changed state; and no sooner is this fountainhead cell formed than it starts to divide into two. This first cell division is a sign of things to come – more and more divisions will follow until a billions-of-cells baby is born.

But why sex?

So the egg has shut the gate and pulled up the drawbridge to the entry of more sperm. Just one sperm has defied the odds and successfully made its way into one egg, and a new, unique life will soon take shape. The story of a pregnancy has begun.

Of course, this strange little dance of invisible cells is how we all begin, but the very otherworldliness of the whole process begs a question that has plagued scientists for thousands of years. Why do we reproduce in this way? Long before the discovery of the microscope allowed us to discover how a sperm fertilises an egg, great thinkers pondered the nature of human reproduction. The male–female duality of human procreation has obsessed mankind for as long as we can tell, and it has formed a central pillar of almost every known belief system in history: Adam and Eve, yin and yang, Aphrodite and Adonis.

The unique ability of sex to generate new life has led many different cultures to link it to the passage of the seasons and the creation of the Universe. In perhaps the oldest religious text in existence, the Aryan-Hindu Rig-Veda, all living things and all other gods are described as springing forth from the sexual union between the Heaven and Earth deities Dyaus and Privithi. Similarly, the ancient Greek pantheon is descended from Mother Earth Gaia and her consort Uranus, the sky (although, true to Greek tragic form, Uranus was subsequently castrated by his own son, who then married his sister, and so on). In chillier northern climes, where the coming of spring must at times have seemed more important than the creation of the Universe, sex was identified as the force that drove the seasons. Probably, Father Christmas is the personification of the Sun, which dips low at the winter solstice to impregnate his lover, Mother Earth, and herald in a new year of fertility. I'll leave it to you to ponder further on the sexual imagery of the Father Christmas myth.

One aspect of sex has been particularly compelling to mankind through the ages. As soon as men and women become fully self-aware adults, their bodies enter an unavoidable and irreversible physical decline towards decrepitude, senility and the humiliation of death. It is almost as if the gods are deliberately taunting us – for the brief years when we are still young and beautiful, we are too immature to appreciate it. Whether they have stated it explicitly or not, most cultures have realised that sex is the only way out of this decline. All we can do as we embark on our terminal deterioration is have children, because producing descendants is the closest we mortals can come to immortality. Thus, sex is compelling to us not just because it is enjoyable, but because it is the only way that we can somehow cheat death.

Of course, today we know much more about the mechanisms of sex, but we can see that the ancients were essentially correct when they saw sex and death as the two forces that control our lives. Until recently, however, we still did not know why sex is the way it is – male and female, sperm and egg. Just because we know how a little bit of a man and a little bit of a woman can make a baby, this still does not explain the central mystery of human sexuality. It does not explain why this is the way that a new person is made.

As we have learnt more about our own bodies and the natural world around us, we have discovered just how much sex is woven into the fabric of life, and how much of our time and energy is devoted to it. Sex is everywhere. Whether you are a biologist or a guardian of public morals, it is hard to deny that sexual reproduction is ubiquitous. In the human world, perhaps a hundred million acts of sexual intercourse take place each day – a thought-provoking statistic if ever there was one – and among animals there are few 'everyday' species that do not perpetuate themselves by male–female union.

Everywhere you look in nature, something is courting, copulating or conceiving. At coral reefs around the world on certain rare nights, the act of procreation is played out in dramatic fashion when crystal-clear seawater becomes thick with the eggs and sperm of tiny reef-building organisms. Sedentary coral polyps indulge in a submarine orgy to produce the mobile jellyfish-like

18

embryos that will disperse to seed new coral colonies elsewhere. No less dramatic are the battles fought by land animals for the chance to procreate. Here the peacock is the rule rather than the exception, as in species after species one sex puts on a display to impress potential mates or intimidate the competition.

Throughout all this frenetic fertility, one thing is clear about sexual reproduction: it is extremely demanding. Much of the energy that animals use when they produce offspring is expended in finding a mate, courting them and making sperm and eggs, many of which will be wasted. The same is true of people and, as any human teenager can tell you, courtship is immensely labour intensive. The apparent waste of time and energy involved in sex is remarkable enough, but when you consider that the efforts of not one but *two* individuals are required to produce a baby, you really do start to wonder if life would be much simpler if we could just spontaneously conceive a perfect replica of ourselves. After all, if this were possible, we could be sure of a guaranteed contribution to the next generation without all those confusing dealings with the opposite sex.

Yet sex remains all-pervasive and, in fact, the development of sexual reproduction was a central event in the story of animal life on Earth. Sex seems to have evolved very early in the line of descent leading to humans. The use of the eggs-and-sperm method of reproduction must be extremely ancient for it to have been inherited by the diverse array of beasts that now use it – almost every major group of animals contains sexual species. Sex must date back at least to the last common ancestor of corals and humans, because we probably both inherited sex from that ancestor. It seems that no sooner had one-celled creatures first clubbed together to make a many-celled animal than some members of the new cellular commune specialised into eggs and sperm, fated to meet and mingle with the sperm and eggs of others.

Despite the preponderance of sex, it is not inescapable. There are other ways of reproducing – asexual ways – and many animals use them successfully. You may remember peering through a school microscope at a single-celled amoeba while your biology teacher assured you that this little creature simply splits down the middle to reproduce itself. Its two daughters are then free to pursue independent amoebic lives. In the next lesson, you may have been presented with the rather more convincing view of a hydra with a baby hydra clearly budding off its side. Despite these examples, binary fission and budding do not somehow seem either practical or convenient for people, or frogs, or insects for that matter. There is, however, another asexual option we could have used, a kind of immaculate conception called parthenogenesis (*parthenos* is Greek for 'virgin'). Parthenogenetic species consist entirely of females that can procreate without the help of males – they can conceive spontaneously.

Although animals could breed by binary fission, budding or parthenogenesis, most reject these options in favour of sexual reproduction. Sex really does seem to be a natural, if not essential, extension of the ability of all living things to reproduce. In the eternal quest to breed, many species have progressed beyond simple self-replication to acquire the ability to produce hybrid offspring between two individuals – sex. But the question remains: why sex? Why have so many animals exchanged the ability to produce a perfect little replica of themselves for a risky and tiresome system of sexual interaction with others? It is unlikely that evolution has endowed us with sexuality just to spice things up – there must be another explanation.

Until the last one hundred years, the reasons why humans use sex to reproduce remained a mystery. Although the microscope had told eighteenth- and nineteenth-century biologists a great deal about how sexual union creates a new life, the 'why' of sex was still unknown. As the nineteenth century drew to a close, all this was to change, largely because of the work of one man. In a series of simple experiments carried out in a small monastery vegetable garden in what is now the Czech Republic, an Augustinian friar was to discover how parents bequeath their characteristics to their children. Although no one realised it at the time, inheritance is the key to understanding sex.

Sex and peas

Born in 1822 in Hyncice in Moravia to a provincial farming family, Gregor Mendel did not seem marked out for a place in history. Although he was a bright boy, his career options seemed rather limited until he became an Augustinian monk in an attempt to gain access to an education. Even then, things did not run smoothly. Although he enrolled at the University of Vienna, young Gregor's intellectual ambitions seemed to fade, and at the age of thirty-one, he retired from academia to become a friar at the Augustinian monastery of St Thomas in what is now Mendel Square in Brno.

Despite what had proved a disheartening foray into the academic establishment, Mendel maintained a strong interest in plant biology. He was especially inspired by the question of how offspring inherit features from their parents. It was obvious to the most casual observer that animals often resemble their parents, and the same is true of flowering plants – breeding two plants that share a distinctive trait, such as height or colour, often results in progeny that also exhibit that trait. Crucially, Mendel strongly believed that these characteristics are passed from generation to generation by simple physical means, and although he was not unique in this belief, he was to be the first to throw light on the subject since the ancient Greeks. The nature of inheritance was very much a matter of debate in Mendel's time. Although philosophers such as Hippocrates had proposed that the essence of a parent's body was in some way physically concentrated into the semen or menstrual blood and that these fluids somehow generate children, many thinkers of the Christian era had maintained that inheritance had an indivisibly spiritual aspect.

There was no evidence to support the notion that parental characteristics were passed on to children by material rather than spiritual means until the advent of the microscope in the seventeenth century. Even then, over-interpretation of the fuzzy images seen through early microscopes led its proponents astray into the erroneous theory of preformation. Supporters of the theory of preformation claimed that either the sperm or the egg contained a complete human baby in miniature, which simply needed to be nurtured to full size in the womb. The characteristics of the child were determined solely by the parent who contributed the preformed micro-baby. Anton van Leeuwenhoek, the Dutch inventor of the microscope, claimed to see a preformed baby in each sperm, whereas Marcello Malpighi, Italian anatomist and sometime physician to Pope Innocent XII, described tiny babies within the substance of eggs.

While these notions may seem strange today, they were widely accepted at the time. Yet, one problem with preformation became clear almost as soon as it was expounded: each tiny baby would have to contain all its preformed children already in its own sperm or eggs. Of course, each of these children would have to contain their own offspring, and those would have to contain their children, and so on. The preformationists could not explain exactly how each sperm or egg could be like a Russian doll, containing an infinite series of ever-smaller descendants, but in the absence of any better explanation, preformation remained popular until the nineteenth century.

When Mendel started to study inheritance, it lay locked in a confused stalemate between its Hippocratic origins, spiritualism and preformation. Back in Brno, unencumbered by the pressures of lab administration and grant applications that confound scientists today, Mendel was free to carry out simple studies of inheritance at his own pace. His scientific genius allowed him to see through a set of apparently trivial results to a simple truth underneath: how children inherit characteristics from both their parents. Instead of searching for the process of physical inheritance of characteristics, Mendel simply assumed that such a process exists and went on to try to deduce how this process distributes parental characteristics to offspring. This intellectual leap from process to outcome was what made Mendel unique among his contemporaries. Although the mechanical process underlying what he discovered would not be elucidated until late in the twentieth century, his work on the outcome of this process was to drive all future studies of inheritance.

Mendel's raw material was nothing more than the humble pea

plant. For his experiments he needed to be able to control how they bred, and he did this by artificially pollinating them. His choice of the pea was probably influenced by the fact that many of the pea plants in the monastery garden exhibited certain obvious, discrete characteristics, such as being tall rather than short, or having red rather than white flowers. It was these simple traits that fascinated Mendel, and they formed the basis of his work.

Mendel soon noticed a simple difference between the offspring of short and tall plants. It became clear that short plants pollinated by other short plants (or by themselves) always produce short offspring. The same was not true of tall plants: the crossing of two tall plants could produce a mixture of tall and short plants. This may not seem a very exciting discovery, but to Mendel it was a challenge. This apparently arbitrary difference between tall and short plants seemed to hint at some underlying mechanism.

Mendel saw that short pea plants were evidently always 'pure' as they could never produce tall offspring – as if there was no 'tallness' within them. In contrast, some tall plants were clearly 'tainted' by a hint of smallness that could find expression in their descendants. Mendel did find, however, that some tall plants were 'pure' in the same way that all short plants were 'pure': they only yielded offspring that were like themselves. Remarkably, this apparently trivial classification of tall pea plants into 'pure' and 'tainted' was the key to a discovery that was to make Mendel one of the most famous names in the history of science.

One thing that Mendel had on his side was time, as you can imagine for a man who spent his life studying crosses between strains of pea plant. He meticulously established crosses between short plants, 'pure' tall plants and 'tainted' tall plants, and the offspring of these crosses were to provide him with his theory of inheritance. Remarkably, several of the crosses yielded tall and short plants in neat mathematical ratios, such as 50 per cent tall and 50 per cent short, or 75 per cent tall and 25 per cent short. These clear-cut proportions must have seemed all the more remarkable to a man working at a time when the process of inheritance was completely mysterious. He seized upon the ratios to infer that they were a sign of a fundamentally simple process controlling inheritance – a process that no one had even been close to discovering before. And he was right.

The solution he devised forms the basis of all modern studies of inheritance. The rules are encouragingly simple. First of all, each individual has two sets of instructions for any given trait (such as height), and it inherits one set from each parent. In turn, this individual will pass on one of these two copies, selected at random, to each of its offspring. These parental instructions are now called 'genes', from *gignomai*, the Greek for 'to be born'.

The only other rule in Mendel's system is that one of the two genes that an individual has for a particular trait may overwhelm the effect of the other. In the case of height in peas, the 'dominant' gene is the one for tallness, and we usually denote dominant genes by capital letters, so we can call this gene T. The gene for shortness is a weaker 'recessive' gene that can be overwhelmed by T and so is called lower-case t. Because of the domination of t by T, individuals that inherit a t gene from one parent and a T gene from the other will be tall: a Tt plant will be indistinguishable from a TT plant. So recessive genes are often hidden, and as we shall see, they are the main reason we have sex.

This may seem like rather a simplistic system to explain something as enigmatic as inheritance, but Mendel could not find a model that explained his results better. The main legacy of his pea plant studies is that they tell us much more than how to breed garden vegetables. Mendelian genetics has turned out to be a very powerful theory because it tells us how characteristics are inherited in all sexual species, and not just peas. The theory will pervade our story of pregnancy, and this is why I have described it in this first chapter, before hormones, birth or even embryos. Mendel's system sums up our modern belief that characteristics are inherited by physical means, and it tells us the simple rules that control that inheritance. Along with Darwin's theory of natural selection, Mendel's genetics have laid the foundations of modern medicine and biology.

Mendel realised that one of the most compelling implications of his theory is that individuals can have a single recessive gene without any external signs of its presence, because it is overwhelmed by a dominant gene from its other parent. These individuals who harbour an invisible recessive gene are called carriers. Recessive genes are found in many inherited human diseases. Single recessive genes float around unseen in the human population until by chance they occur in both partners in a couple, and when this happens problems arise in their children.

Sickle-cell anaemia is one such disease. Most people carry only the normal gene S for haemoglobin, the oxygen-binding red pigment in blood. However, a considerable number of West Africans also carry a 'defective' gene and so are Ss. Because they have one dominant S gene, these carriers are perfectly healthy, but the trouble starts when they have children with a partner who is also a carrier. By the inexorable geometry of Mendel's genetic ratios, each of their children will have a 25 per cent chance of inheriting an s gene from each of its parents. These ss babies have no protective dominant S gene to make normal haemoglobin, and so they show symptoms of sickle-cell disease.

Sickle-cell anaemia is a helpful example of how harmful recessive genes can lurk in the population, often hidden by the curative effects of their dominant counterparts. Each of us carries many of these recessive genes, but most of us feel no ill effects because they are masked by the presence of healthy dominant genes. As long as we do not produce children with a partner who carries the same recessive gene, everything should be fine.

Sickle-cell anaemia also highlights another important aspect of genes: it explains why the gene you inherit from one parent can dominate the gene from your other parent. Mendel could not have explained why the S gene dominates the s gene, but we now know exactly why this is so. The reason that Ss individuals have, rather than do not have, functional haemoglobin is that the S gene itself contains all the information that your blood cells need to make haemoglobin. The gene is not just the switch to turn on haemoglobin production – instead it is, itself, the instructions on how to produce haemoglobin. As long as a blood cell has one S gene, it can produce normal haemoglobin. The s gene can be thought of as simply not carrying the appropriate information to make normal haemoglobin, and so an ss person cannot make any.

This is often why genes are dominant or recessive: the dominant gene carries a functional copy and the recessive gene does not.

At first sight, Mendelian genetics, with its neat little genes fluttering down through the generations and precisely dictating our characteristics, does not really seem to explain all the subtleties of inheritance. For example, although short human parents often produce short children, they do not always do so. Indeed, many characteristics, such as height, weight and intelligence, seem to be inherited in a much more vague way than the clear-cut examples of pea plant height and sickle-cell anaemia. However, it now seems that Mendel was right all along – his genetics can explain the intricacies of inheritance. The only reason that inheritance seems an imprecise process is that many traits, such as human height, are controlled by not one, but many genes. These genes are inherited in exactly the way Mendel predicted, but because there are so many of them, their cumulative effects are much harder to predict. In fact, it is very lucky that Mendel studied traits in peas that are inherited in the form of single genes, or he would probably have been overwhelmed by the complexity of his results.

Gregor Mendel's system of genetics has become so ingrained into the modern mind that, rather like Darwin's theories, it is easy to overlook just what a unique discovery it actually was. Mendel's genetics marked a new direction in the study of our sexuality, as scientists suddenly had a theory of inheritance that was intimately tied up with sex. In fact, Mendelian inheritance and sex almost seemed to be two sides of the same coin. Sex allowed inheritance to occur, and inheritance looked like it might contain the explanation to the ancient question of why sex exists. Mendel's theory was immensely powerful, in a way that only such a simple theory can be. Indeed, the British evolutionary biologist Sir Gavin de Beer elevated Mendel's achievements above all others: 'There is not known another example of a science which sprang fully formed from the brain of one man.'

Yet, Mendelian genetics was certainly not hailed in these terms at the time of its publication. Mendel himself clearly underestimated the significance of his work when he published 'Versuche über Pflanzenhybriden' (Experiments on Plant

Hybridisation) in the rather provincial *Journal of the Brno Natural History Society* in 1866. No doubt this is one reason why the theory languished virtually unread until the start of the twentieth century. Another is that many people simply did not understand why the theory was so important.

Mendel was not embittered by the lack of reaction of the scientific community to his ideas. He continued to pursue his quietly learned existence for the rest of his years, and rose to the position of abbot in 1868, two years after the publication of the 'Versuche'. He died on 6 January 1884, but the diversity of his interests and achievements was clear in the elegiac description released by his monastery on the day of his death:

Incumbent of the Prelature of the Imperial and Royal Austrian Order of Emperor Franz Josef, meritorious director of the Moravian Mortgage Bank, founder of the Austrian Meteorological Association, member of the Royal and Imperial Moravian and Silesian Society for the Furtherance of Agriculture, Natural Sciences and Knowledge of the Country, and of other learned and beneficial societies.

Mendel's story is often quoted as a rather romantic example of scientific inquiry: an insightful savant, equipped with nothing more than an inquiring mind, trying to explain the world around him. However, recent re-analysis of his results has rather spoilt our idyllic image of Mendel. It is not that his results were wrong in any way – his experiments have been repeated often enough to show that this was not the case. Quite the opposite, his results may have been too good. When contemporary scientists carry out an experiment similar to Mendel's, they do not expect his 50%:50% or 75%:25% ratios to be generated perfectly, because nature simply does not behave like that. Nature is full of random variation, and 47%:53% or 77%:23% is as close as studies in biology are likely to approach perfection. It has been calculated that the odds against Mendel obtaining his amazingly exact ratios are enormous. This is rather a surprise, as it implies that Mendel did not quite play by the usual scientific rules. Perhaps he counted pea plants until he obtained a convincingly exact ratio, or perhaps he discounted a few trials that 'did not work' as hoped. It has also been suggested that Mendel had actually expected a completely different result and was desperate to salvage something from all his work. We will probably never know exactly how he produced his suspiciously good results, but somehow the thought of this infuriated Augustinian friar scribbling over results that simply wouldn't cooperate is a rather endearing one.

Why sex is more than just fun

What does Mendel's genetics tell us about why so many animals reproduce using sex? First of all, Mendelian genetics confirms that sex is a compromise by an individual keen to pass its genes on to the next generation. Rather than being able to pass on all your genes to your children, Mendel's system dictates that you may pass on only half of them, allowing the other half to be discarded and replaced by those of your mate. Worse, you have no control over which half of your genes is passed on and which half is thrown away. Admittedly, you can pass on more of your cherished genes by having more children, but this is an extremely demanding business, and having too many children could jeopardise your ability to look after any of them properly. Anyway, the fixed laws of genetics dictate that no matter how many offspring you had, you would never be able to pass on all your genes. Why do we have to accept this state of affairs?

Sex is so popular because things change. We animals are not living in a carefully tended vegetable garden and there is no benevolent guiding force protecting us. Just as people gradually and irretrievably deteriorate as they get older, so do our genetic instructions succumb to wear and tear over the generations. Our genes are remarkably stable over time (scientists can retrieve gene fragments from dinosaur bones), but they do get damaged. A noxious chemical here or a random cosmic particle there and we are left with irreversible damage to our genes. These sporadic genetic injuries – or mutations – cannot always be repaired. Mutations change our genetic instructions and, in doing so, usually render them useless. Just as a mistyped word in a book usually makes the book less easy to understand, so do most mutations adversely affect the function of a gene. Mutation is most threatening when it takes place in our germ cells – the sperm and eggs that will produce our offspring – as these mutations will be faithfully reproduced into every cell in your children's bodies. Obviously, animals must have a way of coping with this continual genetic damage if they are not to deteriorate into extinction.

Our short-term solution to the problem of genetic damage is to have two sets of our genetic instructions (we get one set from each parent), because if a particularly useful gene gets damaged, we still have a good copy to fall back on. Unfortunately, without sex, carrying around a duplicate set of all our genes is only a temporary solution to our problems because, as time passes, our descendants could eventually suffer a mutation in the remaining intact copy of the gene and be left with no functional gene at all. Our progeny would gradually accumulate mutations in both copies of lots of different genes and have no way of getting rid of them. What seemed to be an economical system would lead to disaster.

The best way to stop harmful mutations accumulating seems to be sex: intimate genetic interaction with other individuals prepared to cooperate. If you produce a child with a partner you are doing exactly this: you are giving your child half of your genes along with half of your mate's genes, just as Mendel would have predicted. When, in turn, your child produces a grandchild, about one-quarter of your genetic material will be passed on. Similarly, about an eighth of your genes will end up in each great-grandchild.

Now comes the clever part. If a human population is neither growing nor shrinking, then each couple has an average of two children. Both of these children inherit half of each parent's genes. By the same logic, Mr or Mrs Average also has four (2×2) grandchildren, each with about a quarter of their genes. The story then continues: eight ($2 \times 2 \times 2$) great-grandchildren, each with about an eighth of their genes; sixteen ($2 \times 2 \times 2 \times 2$) great-great-grandchildren, each with a sixteenth of their genes. In other words, each subsequent generation contains roughly one set of each ancestor's genes. The only difference that sex makes is that this set of genes is now spread over many individuals, rather than being concentrated in a single narrow line of descendants. So sex perpetuates our genes at the expense of dispersing them through many offspring. But is this dispersal such a bad thing? After all, your functioning 'good' genes are randomly passed on to different progeny from your damaged ones. The lucky offspring can flourish, but a few unfortunates get too many damaged genes and die out. In this way, the population to which you contribute can gradually discard more 'bad' genes than 'good' ones.

In fact, your genes will be spread extremely thinly among an enormous number of descendants much more rapidly than you probably think. Imagine that William the Conqueror produced two children around 1050. Then imagine that each of these produced two children twenty-five years later. Finally think of this process happening every twenty-five years until 1975. This covers a span of 925 years, which is thirty-seven generations. In an infinite world, this would mean that William would have 137,438,953,472 descendants today. Clearly this is silly because 137,438,953,472 is around twenty times larger than the current world population – his actual number of descendants is actually far less, because they are trapped on a small island in the Atlantic and have been forced to breed with each other. However, the fact remains that William's genes will now be spread extremely thinly over a huge number of people.

So if we look at a period as short as one millennium, each human individual is exposed as an evanescent being: a repository for a minuscule fraction of the enormous genetic soup that swirls through the human population. Sexual reproduction means that our effective legacy is our dissociated genes, scattered indiscriminately into future generations.

This is probably the main reason why we have sex. Sex stirs the genetic pot and constantly mixes human genes into new combinations that make new people. People unlucky enough to get lots of damaged genes are discarded and the bad genes are lost from the race for ever. Sex means that the human race is not a neat arrangement of family lines, but a heaving mass of genes that is able to defend itself against its mutation-causing enemies.

Other reasons to have sex

So, for a population of animals that is neither growing nor shrinking, and which lives in a nice unchanging environment, sex allows harmful genetic mutations to be discarded.

Yet organisms' environments do change. The climate alters, food becomes more or less scarce, diseases wax and wane, and competitors come and go. How does a population of animals respond to these challenges, and how can sex help? Because we sexual species spend all our time mixing our genes with each other, we are a pretty varied bunch. No two sexual animals are exactly the same, and most of us differ in quite striking ways. Because of this variation, when the environment changes, some individuals cope better than others. These individuals then usually go on to produce more offspring as a result of their success. This is useful because these animals have coped better because of the mix of genes they carry, and when they breed, these helpful genes are passed to the next generation. Thus, sex is an excellent way for an animal population to adapt to environmental change, because it helps to propagate exactly those genes that allow animals to survive in the new environment.

Sometimes something radical and rare also helps animals to adapt to a changing world. Occasionally a gene is mutated in a way that, rather than damaging it like most mutations, makes it more useful in the new environment. Once again, sex allows this new gene to spread quickly through the population, where it can mingle with other helpful genes and aid the species' adaptation to new circumstances.

Sex is not just a way of spreading or removing genes – it helps animals to fight disease. It does this by allowing animals to make the most of a special set of genes – the major Histocompatibility Complex or MHC – which is an integral part of the body's defences against infection. For reasons that will become clear in Chapter 4, individuals that inherit a different set of MHC genes from each parent are probably able to fend off infections more effectively. Obviously, an animal can only inherit different genes from two parents if it is generated sexually.

The final advantage of sexual reproduction is perhaps the weirdest of all: sex may help animals to deter parasites from attacking them. Most animals are constantly locked in battle with a range of creatures that are trying to live inside them, and they have to keep these parasitic invaders at bay if they are to survive. As animals evolve to succeed in changing environments, their parasites have to evolve with them, and it is likely that sexual reproduction makes it harder for the parasites to do this. For one thing, sex allows animals to evolve much faster, so that parasites may simply find it more difficult to keep up. Also, members of a sexual species tend to be quite varied, and this variety also makes it harder for parasites to spread from one individual to another. Sex probably makes it harder for parasites to spread from parents to offspring as well, and it has even been suggested that males were invented to deter parasites that specialised in spreading from mothers to their daughters.

So since Mendel explained how we use genes to inherit characteristics from our parents, we have learnt that sex helps us in several different ways – removing damaged genes, disseminating helpful ones and protecting us against disease. Sex is an essential part of our life because it allows us to cope with all the challenges that the world can throw at us. The old idea that sex is a way of cheating death has turned out to be true in a far deeper way than we could ever have imagined. Humans cannot be immortal because their genes would gradually deteriorate to a point where they could no longer function. Instead, we have sex and die so that our genes can live forever.

Life without sex

If sex is so useful, then how do asexual species cope without it? Sexual reproduction is obviously not essential because many animals manage perfectly well without it, but one thing is clear: asexuality is rare among our close relatives.

My definition of 'close relatives' may be rather wider than most people's. Humans are vertebrates, like cats, birds, snakes, frogs and fish, and all vertebrates share the same basic body plan and body

chemistry. Asexual reproduction is very unusual among vertebrates and, in fact, the only asexual land-living vertebrates are
lizards. It is not known why asexual vertebrates are so few and far
between − we do not know why there are no asexual fish, frogs,
birds or mammals. However, lizards seem to have a penchant for
becoming asexual, and there are several different asexual lizard
species in existence, all born by the virgin birth of parthenogenesis.

Here, we are entering a strange ultra-matriarchal world. The
family life of parthenogenetic lizards simply does not seem like
anything we are used to − all the animals that we usually deal with
have well-adjusted sexual families just like us. For a start,
parthenogenetic lizard populations do not have any males at all.
These lizards do not produce any male offspring − it is almost as if
they know that they do not need them. Instead they produce
female offspring, and these produce female offspring and so on, *ad
nauseam*. Instead of a network of interlinked families sharing their
characteristics like we do, parthenogenetic lizard populations
consist of a number of discrete lines of descent, each of which
contains a large number of essentially identical individuals. Because
there is no sexual genetic shuffling going on, each lizard is almost a
perfect replica of its mother, its sisters and its offspring. These lines
of descent are 'clones' in the same way that Dolly the sheep was a
clone of her 'mother' (the sheep that provided her genes).

We have only recently started to resolve the question of where
these parthenogenetic lizard species come from, and the findings so
far are surprising. It is now thought that parthenogenetic females
are initially generated by mating of males and females from two
different sexual species. Despite this unhealthy start, new parthenogenetic species must have some advantage over their sexual
relatives or they would not survive. You might expect that asexual
lizards could save energy by avoiding all that costly courtship and
copulation, but it seems that many parthenogenetic lizards still
court and mate with males of closely related species, although the
sperm they acquire plays no part in the generation of offspring.
They may be parthenogenetic, but they are not celibate.

When they first appear, parthenogenetic lizards carry two sets of
genetic instructions, just like we sexual animals. Yet once these

lizards get established, they often start to produce a new type of daughter with three sets of genetic material, probably because they add a set of genes by mating with males from one of the 'founder' species. Now, just as the original parthenogenetic lizards pro-liferated at the expense of their sexual relatives, the new three-gene-copy animals soon start to edge out their two-gene-copy parthenogenetic ancestors.

With all this inter-species mating and virgin birth, one thing is clear. The world of parthenogenesis is truly bizarre from our own perspective – a tangled web of deviance, abstinence and indul-gence worthy of the *Jerry Springer Show*. Yet the very unusualness of parthenogenesis has made biologists question how these animals fit into the story of life on Earth. Unfortunately for the parthenogenetic lizards, the news is not good. They are a dead end to evolution: they do not seem to go anywhere. Genetic analysis of asexual lizards suggests that they are all very young species – most of them have appeared within the last few thousand years. That there are no ancient parthenogenetic vertebrates alive today strongly suggests that they do not last very long. Therein lies a clue to their fatal flaw. Although they may initially benefit from their celibate existence, their inability to remove damaged genes and adapt to a changing world consigns them to rapid extinction. Parthenogenetic species are transient offshoots from sexual species that prosper briefly, but soon enter a terminal decline. They simply cannot cope as well as their more adaptable sexual relatives.

But sometimes asexual animals do not have to be evolutionary dead ends. Although there are no lizards that can alternate between parthenogenetic and sexual lifestyles, some non-vertebrates do seem to have the best of these two worlds. Hold a jar of pond water up to the light and one of the tiny creatures you will see darting about will be *Daphnia*, the water-flea. Water-fleas have two different life cycles, one sexual and one asexual. Populations of sexual water-fleas occasionally generate several asexual individuals and these flourish at the expense of their sexual colleagues until their descendants eventually revert to a sexual lifestyle. These new sexual animals then re-establish the genetic variation present in the original population. By constantly switching between sexual and asexual forms, the water-flea presumably gains the advantages of both lifestyles – asexual proliferation and sexual rediversification. So parthenogenesis does not have to be a dead end, especially if animals can revert to their old sexual ways when they need to.

Human reproduction without sex?

These lizards may seem very different from ourselves, but in the great scheme of things they are very closely related to us. The genetic and anatomical scheme of the lizards that gave rise to these exceptional species is very similar to our own, so there is no simple reason why they are the parthenogenetic ones, and we are not. For some time, scientists have wondered if there is a mechanism that prevents women from spawning a new race of parthenogenetic people.

This idea has surprising parallels in mythology. One of the products of the exuberant promiscuity of the ancient Greek pantheon was the Amazons, an all-female race of hunters who lived in Thrace (or the Caucasus, depending on who you read). Like parthenogenetic lizards, the Amazons were generated by an inter-species mating (between a god and a nymph) and they also displaced the mixed-sex races from the areas they inhabited. They were not, however, truly parthenogenetic as they had to mate with men, and produced male children (whom they killed, of course). This may be a rather flippant example, but the same cannot be said of ancient musings on virgin birth. When the early Christians wished to infuse Jesus' otherwise mundane early life with a hint of divinity, they claimed that he was produced by virgin birth. Virgin birth was so obviously impossible in the normal course of events that this plot device was enough to set the Messiah apart from all those run-of-the-mill prophets. Some Christian sects have also cleverly picked up on the point that this puts his mother in an interesting spiritual position. Mary was an integral part of this divine parthenogenesis, so what does that make her?

Some evolutionary biologists have found it rather surprising that there are no parthenogenetic species among our closest relatives, the mammals. They argue that parthenogenesis may actually be favoured in species where eggs are fertilised inside the female and embryos are nurtured internally. Potentially, internal incubation of embryos could allow asexual mothers to discard genetically damaged offspring before they are born, so postponing the genetic 'day of reckoning' that threatens all parthenogenetic species. If this theory is right, then parthenogenesis should be more common in live-bearing mammals than egg-laying reptiles. Yet as we have seen, this is clearly not the case.

In fact, it is now thought that parthenogenesis is actively prevented in mammals, including ourselves. Studies of early embryos have shown that genes inherited from the father are necessary for normal formation of the placenta, the organ that keeps the fetus alive by allowing it to draw nutrients from its mother's blood. Somehow the embryo knows that these genes came from the sperm rather than the egg, and will use them only if they came from the sperm – a process called genetic imprinting. Because of this, an embryo that did not have a sperm involved in its formation cannot make a placenta and so cannot be born.

The unusual dependence of the placenta on paternal genes has given mammals a neat system for preventing human parthenoge-netic babies. Whether this system evolved specifically to prevent mammalian parthenogenesis is unclear, but even if it has developed for some other reason, it is hard to see how it could allow parthenogenetic humans. Even though unfertilised eggs can start to divide, and can develop for over a week, the resulting all-maternal embryos can never survive until birth without a placenta. Immaculate conception may happen, but virgin birth is out of the question.

A recent chance discovery has confirmed the need for paternal input in human pregnancy. In 1995, a research team in Edinburgh was studying a one-year-old boy, 'F.D.', with an unusual set of abnormalities. The left side of his face is slightly smaller than the right side, his uvula is split into two (the uvula is the dangly thing that hangs down at the back of your throat) and he has mild learning problems. When the researchers studied the genes in a blood sample from the boy, they found that he apparently had the genetic constitution of a girl. This is an unusual, but not extremely rare, occurrence, and so the team thought that the boy was a case of 'sex reversal', a condition where the normal mechanisms that control a child's sex fail to operate, as we will see in Chapter 3. Remarkably, however, when they studied the genes in some skin from his leg, they found that these cells were genetically male. This raised the possibility that he was a mixture of male and female cells, a very rare individual called a chimaera (named, rather indelicately, after a mythological monster made up from parts of different animals). Yet, when they compared the genes in his leg with those in his blood, they discovered to their surprise that all the cells in his body had come from the same egg. This led the scientists to the conclusion that F.D. was, in fact, unique. Half of his cells are parthenogenetic. He is a male/parthenogenote chimaera.

The team then pieced together the first few unusual days of F.D.'s existence. His mother produced an egg, and this egg misguidedly started to divide before it had been fertilised – it split into two. Then something extremely unusual took place. A sperm fought its way into the zona pellucida and fertilised one of the two cells. The fertilised cell then started to divide and produced the half of the boy that is 'normal', and the unfertilised cell also began to divide and ended up producing the half of the boy that is parthenogenetic. So, F.D. is half boy, half parthenogenote. For example, the left side of his face may be smaller because it contains more parthenogenetic cells – they appear to grow slightly less well than normal cells. Obviously, F.D. must have made a placenta, or he would not be with us, so a lot of paternal-gene-containing 'boy cells' must have made it into the placenta. Unfortunately, F.D.'s placenta was discarded long before anyone realised how special he is. So, F.D. proves that human parthenogenotes can survive, but only if they are mingled with sexually generated cells.

Luckily for F.D, there appear to be quite a few 'boy cells' in his testicles as well, because he is thought to have a normal reproductive system, and the researchers suspect that he will be fertile and will be able to father normal children.

Our hunt through the weird world of asexuality has confirmed what we had suspected. Sexual reproduction is the best long-term plan for the health of our genes. There is a tremendous genetic imperative to mingle our genes with other people's, and this is probably why so many of us spend much of our time trying to attract the opposite sex.

We have also stumbled on a disquieting fact. I suspect that most people believe that mothers and fathers contribute equally to the production of their children, genetically at least. The Mendelian idea of equal contribution by both parents certainly appeals to our modern concepts of equality, but the discovery of genetic imprinting has thrown a spanner in the works. The fact that the embryo needs paternal and not maternal genes to make a placenta contradicts our ideas that there is a symmetry between the sexes. We will soon see that this symmetry is, in fact, completely illusory. Males and females do not contribute equally to pregnancy. Aside from the need for paternal input to make a placenta, it is your mother who dominates your early existence, and she does so in rather profound ways. Your mother, it seems, knows best.

Why are eggs big and sperm small?

In the same way that people have always wanted to explain the existence of sex, we are also obsessed with the differences between the sexes. It is no accident that most of our literature can be seen as a continuing attempt to explain why men and women treat each other the way they do. Of course, the two philosophical questions of the existence of sex and the nature of the sexes are very closely linked. Sex is the union of two different sexes, and so the sexes must be different. But why do they differ in the ways they do? It now seems that we have the answer to this ancient question. The different roles that men and women play in creating their children all seem to stem from a single biological difference between them: eggs are much bigger than sperm.

While I was working at the Institute of Zoology in London, my research group was interested in developing techniques that we could use for breeding endangered species for reintroduction into the wild. As part of this, I was trying to develop methods of *in vitro* fertilisation (IVF) for deer. This turned out to be a distinctly fiddly process. We collected eggs from females by laparoscopy, just like doctors do in human IVF clinics, and we also collected sperm from stags (extremely carefully). These germ cells were then washed and prepared for fertilisation. The sperm and eggs were then mixed in a plastic dish and I usually had a few seconds to watch them before returning them to their warm incubator. The day's labour had often put me in a rather philosophical mood by this time, and as I watched the sperm swarm into view and crowd towards the waiting eggs, a strange question entered my mind. Why are eggs so big and sperm so small?

Eggs are huge for single cells – around a tenth of a millimetre. They are clearly visible to the naked eye and larger than almost every other cell in the human body. Eggs are not only large, but also highly complex – factories packed with intricate machinery ready to start manufacturing a baby. Indeed, eggs are so geared up to making embryos that they can even accidentally embark on the first few rounds of embryonic cell division without ever having seen a sperm.

In contrast to eggs, sperm are distinctly tiny, even by the standards of cells. They consist of no more than the minimum they can get away with: genes crammed into the sperm head, a tail driven by a compact biological outboard motor and some tools needed to dig their way into the egg. Even their genetic material is packed extremely tightly, far more so than in any other cell. Sperm's genes are packed so tightly, in fact, that they do not seem to be able to use them. Because of this, much of the construction of mature sperm is carried out not by their own cellular machinery, but instead by helper cells in the testicles.

I began to wonder why there is such a disparity between the size and complexity of male and female sex cells. I was reassured to find that I was not the only person who had ever wondered about this, and I found a small volume of literature published on the subject, mainly by evolutionary biologists. The term they have invented for the sperm/egg size discrepancy is 'anisogamy', literally 'unequal marriage'. Anisogamy is one of the facts of life that we all take for granted: eggs are big and sperm are small, but we do not know if things have to be this way, or why.

At first I found anisogamy interesting simply because it seemed like an apparently arbitrary fact demanding explanation. However, I soon began to wonder if it was the thing that made the sexes different. After all, in most animals, the sex that makes the eggs is usually the sex that puts most of the effort into nurturing offspring. Of course, in a few species the male does all the work, such as the pregnant male seahorse or the male emu dutifully rearing his chicks. However, male investment in childrearing is not a frequent sight in the natural world. Females seem to pay twice − they put lots of energy into making their large eggs, and then they have to protect their investment by nurturing them once they are fertilised. Males can have a much more cavalier attitude − their sperm are small and cheap, and their interests are often better served by going off and fertilising someone else.

So anisogamy may explain a great deal about the differences between the sexes. It could even answer the question often screamed by women during labour: why don't men have to give birth?

An obvious alternative to anisogamy would be 'isogamy': males and females producing sex cells of roughly the same size. Isogamy turns out to be surprisingly uncommon in the animal kingdom, however, whether or not animals' eggs are fertilised within the female. There is little difference between corals that release their germ cells to float off and meet in the open ocean, sticklebacks where the male squirts sperm directly on to the female's freshly laid eggs and humans where fertilisation takes place discreetly inside the female. In all three, sperm are small, superabundant and usually wasted, whereas eggs are large, few and precious.

There are several possible reasons why anisogamy is so common, and the first is perhaps the simplest of all: sperm is small so it can be mobile. Indeed, sperm are usually much more mobile than eggs and this mobility is often important in ensuring that the two meet. Because of this, it has been suggested that eggs and sperm are different sizes so that one can hunt out the other. Perhaps the immobile cell can be as large as it likes because all it has to do is sit and wait. In fact, it is likely that it would be beneficial for the egg to be as large as possible, since it will contribute the bulk of the embryo.

The need for one sex to produce mobile germ cells is a

reasonable explanation for anisogamy, but it is unlikely to be the whole answer. For one thing, sperm mobility is unlikely to be such a universally important factor that it has led to huge eggs and tiny sperm in so many animals. For example, sperm mobility is much less important in the many animal species in which sperm are ejaculated directly on to the eggs. Another important issue is whether a small sperm is more likely to reach the egg than a big one. If human sperm were the size of tadpoles, then there is probably no reason why they should not be able swim like tadpoles. And, of course, a tadpole can swim the distance equivalent to that between the vagina and the Fallopian tube considerably faster than a sperm.

Another possible reason for anisogamy could be competition between sperm – the idea that sperm are involved in a competitive race to reach the egg and that the sperm that wins the race carries a better selection of genes. One possibility is that men produce many sperm so that the very 'best' sperm can be selected to fertilise the egg. An alternative possibility is that men produce as many sperm as possible to improve their chances of fertilising the eggs of women who have mated with several competing partners. This competitiveness may explain why there have to be as many sperm in an ejaculate as there are people in the United States. And, of course, the best way to produce more sperm is to make them as small as possible.

There are, however, reasons why sperm competition is unlikely to be the cause of anisogamy. We have already seen that there is no reason to believe that small sperm can swim especially fast, so it is difficult to see why small sperm should have any competitive edge over larger sperm. If a male mates a female that has previously been impregnated with the tiny sperm of another male, his interests might be better served by releasing a few tadpole-sized sperm that can overtake their tiny adversaries and fertilise the egg. Indeed, there is evidence that longer sperm are better at out-competing sperm from other males. It is still possible, however, that inter-male competition drives males to produce more sperm per ejaculation, and this may tend to make them reduce the size of each sperm.

There is yet another problem with the competitiveness of sperm: it may not be very important in humans. Humans are probably naturally less promiscuous than many mammals, and so our mechanisms to make sperm compete with those of other males are likely to be poorly developed. There are also problems with the idea that one man's sperm all compete with each other – there is little evidence that sperm that swim well carry 'better' genes. Of course, sperm do vary in their ability to swim because if they did not, hundreds of millions of sperm would reach the egg at the same time. Yet this variation in swimming ability is probably nothing to do with their genes. As we have seen, sperm hardly use their genes at all in transit, and much of the construction of a sperm is actually carried out by other cells, using the genes of those cells. Anyway, even if sperm competition does occur, one could argue that it would favour sperm with good genes for making sperm, rather than genes that are particularly useful for anything else.

Whether or not sperm mobility and competition are the reasons why sperm are so small, they could certainly explain why eggs are so big. Eggs are just as important in this argument as sperm. With the onus of ensuring that the two cells meet placed firmly on the sperm, there seems to be no reason at all why the egg should be at all agile. Freed from the need to hunt out its mate, there are few reasons why an egg should be small, and very good reasons why it should be large. The larger an egg is, the larger the embryo it will eventually produce, and larger embryos will have a head start over their smaller peers.

Even though we may not have yet entirely explained anisogamy, it is becoming clear that there may be many advantages in having a sexual system that is polarised between one sex that makes smaller, mobile sex cells and another sex that makes larger, immobile cells that contribute most of the embryo's resources. This is a challenging idea because it may explain why males and females are so very different. Eggs and sperm can be seen as diametrically opposed strategies for generating children – made by two sexes that optimise their chances of producing offspring in opposite but complementary ways. Different things are important for males and females: males try to fertilise as many eggs as possible and females try to ensure that the eggs they produce give their offspring the best possible start in life. Small sex cells and large sex cells, increasing the number of offspring and increasing the chances that those offspring will be successful – these may be the differences that define what men and women are.

There is one more theory that may also help to explain anisogamy, and it is the theory that has most hard evidence to support it. It is also the strangest theory of all, as it touches on the very question of what we animals are.

Most living things can be classified into two groups. The first group, the bacteria, is by far the largest. Bacteria are tiny bags full of salty water containing genetic material and everything else they need to make them work. The second group of living things is called the eukaryotes ('well-nucleated') and it includes animals, fungi and plants. The distinguishing feature of eukaryotes is that their cells are partitioned into two components: the nucleus, which acts as a repository for genetic material, and the cytoplasm, which contains all the rest of the machinery that runs the cell. The nucleus evolved in the dim and distant past, perhaps one billion years ago, and the ancestral eukaryote eventually gave rise to animals, fungi and plants – you, your cat, mushrooms and sweetcorn.

When the first eukaryotes appeared, they had a problem. While they could break down their food to generate energy, they could not make sugars react with oxygen. 'Burning' sugar in this way is a very efficient way to produce energy, and so the early eukaryotes were at a considerable disadvantage. Their solution was a radical one – they allowed themselves to be infested by small bacteria that could burn the sugar for them. These invading bacteria benefited from the physical protection of their hosts, and of course the host eukaryote benefited from its new-found ability to react sugar with oxygen. To this day, these bacteria persist in just about every cell in your body, and we call them mitochondria. Although they have changed over the aeons, their job remains the same, burning sugars with oxygen, and we could not survive without them.

This mitochondrial infestation provides an important clue to the disparity between the sizes of sperm and eggs. When the ancestral mitochondria invaded eukaryotes, they brought their own genetic material with them. Just like any successful parasite, they insisted on reproducing independently of their host by passing on their genes to their offspring. As it happens, mitochondria do not reproduce sexually but by simple asexual binary fission – splitting in two like an amoeba – but the fact remains that they are there, in all of us, breeding. (Things are more complicated in plants, because they were invaded by a second set of bacteria – green ones that make sugar from air, water and sunlight, now known as chloroplasts.)

Each human individual is, therefore, not one but two creatures coexisting together – our sexual nucleated eukaryotic cells and our asexual bacteria-like mitochondrial guests. Every time one of our cells divides, not only does the nucleus divide, but the mitochondria also replicate themselves and their own genes.

This essential eukaryote–mitochondrion duality of humans may be the cause of the size disparity between sperm and eggs. Imagine if men and women produced sex cells of equal sizes, each carrying its own complement of mitochondrial passengers. After copulation, these cells would seek each other out and fuse, just as sperm and eggs do, and their nuclear genes would mingle. The problem would come when the two sets of mitochondria met in the same cell. They are full-blooded, autonomously reproducing organisms, so there is no reason why they should not try to compete with each other. The maternal and paternal mitochondria would compete just like any other organisms do – by trying to out-populate each other, leading to a predominance of one strain in the developing embryo and its future offspring. An uncontrolled struggle for supremacy between parental mitochondrial factions is of little advantage to the eukaryotic host. Indeed, it could be extremely harmful. After all, the developing embryo has enough to do without having to act as a mitochondrial battlefield as well. The eukaryote needs mitochondria, but not their petty squabbles.

Eukaryotes have cleverly avoided this conflict, by allowing only one parent to contribute mitochondria to offspring. You, for example, received all your mitochondria from your mother and none from your father. This means that mitochondrial genes are a glaring exception to Mendelian genetics, because males' mitochondrial genes are never inherited. There are mitochondria in the sperm, but they are not allowed to become part of the embryo. Maybe this is the most accurate definition of the male and female sexes: males do not pass on their mitochondria to their offspring, and females do. Males are males because of this fundamental self-sacrifice for the common eukaryotic good.

So, anisogamy may be the way we avoid inter-mitochondrial strife. We each inherit our mitochondria from just one parent, and this is the parent who produces the largest sex cell – our mother. Because of her initial investment in the egg, and because she is often much nearer to the eggs when they are fertilised, she then usually has to do the lion's share of the nurturing – either inside or outside her body. By developing this system, we eukaryotes have consigned our bacterial guests to a rather hermit-like future, destined to reproduce for ever isolated from their kin in the privacy of their own host. We have achieved this by preventing the mitochondria from one parent entering the embryo, and in doing so we have also defined our own sexuality.

Mother takes over

Within a day of the sperm entering the egg, the embryo rushes headlong into a frenetic round of cell division, producing four, eight, sixteen cells within three days. Despite all this purposeful activity, the embryo is not really in control of itself. It is not even able to use its own genes, and so it is flying on autopilot, using a set of instructions pre-packaged into the egg long before the sperm arrived. Only a few hours ago, the egg was an integral part of the mother's body and it seems as if it is difficult for her to relinquish control now it has been fertilised.

Just as anisogamy means that your mother had more biological responsibilities than your father, your mother was also able to control the first week of your existence in a way that no one else will ever be able to control you again. The inability of embryos to use their genes during that first week makes them incredibly vulnerable, so their mothers step in and take over for a while. All that frantic cell division is orchestrated not by the embryo, but by a control system implanted into the egg long before it was fertilised.

Genes are made of DNA (deoxyribonucleic acid), a long linear chain-like molecule containing genetic instructions arranged sequentially along its length, rather like an old-fashioned ticker tape. When a cell wants to use a particular gene to make something for itself, it transcribes a temporary copy of its information on to a smaller linear molecule called RNA (ribonucleic acid). This is often called 'messenger RNA' because it migrates out of the nucleus, carrying the information from the gene with it. Often, this messenger RNA is then used to make proteins, which make up most of the machinery that runs a cell.

This is what happens in most cells, but the egg is unusual in that an array of messenger RNA and protein is packaged into it while it is still in the ovary. These RNAs and proteins are not used, however, until the egg has been fertilised some days later. They then form the control system that drives the first few days of embryonic cell division. After that time the RNA and protein start to deteriorate, and there are no maternal genes to replace them – what remains of the mother's genes is now completely mixed in with the genes from the sperm – and so the embryo has to start using its own genes.

This changeover between the use of maternal and embryonic genes is a critical time for the embryo. Mouse embryos switch from maternal to embryonic control at the two-cell stage, humans at four or eight cells, and rabbit embryos not until sixteen cells. In my experience, this transition stage seems to be a very difficult time for embryos – many of them simply do not seem to be able to make the switch, and this was the time when most of my deer embryos died.

Perhaps fertilisation cannot be considered complete until the embryo's own genes switch on and start acting in concert. As we have seen, this genetic autonomy can come surprisingly late. The control that mothers exert over their embryos is considerable, and certainly gives them a good opportunity to further influence the future of the embryo. Maternal messages remaining from the time of the egg do seem to be essential for formation of normal offspring in almost all animals studied, and mothers can use this influence to overwhelm the contribution of the father in striking ways.

Some of the most dramatic evidence of maternal control of the embryo comes from the fruit fly *Drosophila*. If you have read anything at all about biology in the last few years, you cannot have avoided this disgusting little fellow. Fruit flies are a good example of maternal control, as their maternal signals seem to go as far as initiating the formation of the offspring's entire body plan. Pioneering studies of abnormal fly development carried out in the 1980s identified many mutant strains of *Drosophila* with strange defects. Some of these defects struck their discoverers as interesting because they looked as if they were caused by failures of important stages in the formation of the body. In one of the most bizarre mutations, the front end of the fly larva failed to form normally, yielding a maggot with two hind ends. After a considerable amount of work, this defect was found to be due to damage to a single gene called *bicoid*. Unexpectedly, full-blown *bicoid*-mutant larvae arise not when the larva is deficient in the normal *bicoid* gene, but when the mother fly is deficient in the gene. Why should larval development depend more on the mother's genetic constitution than on that of the larva itself? The answer is that the protein made by *bicoid* is made by the mother, not the embryo. *Bicoid* messenger RNA is made by maternal cells that sit next to the egg when it is still inside the mother fly's body. The *bicoid* messenger RNA is then inserted into one end of the oblong egg and is translated into *bicoid* protein. This maternal protein is what confers 'frontness' on that end of the egg. All subsequent specialisation of the front and hind ends of the maggot is critically dependent on this maternal signal, and without it a normal fly cannot be made.

This seems an astounding concession for the embryo to make. It has relinquished all control over establishing the orientation of its own body to its mother. Any prospective fruit fly father had better hope his intended mate's *bicoid* genes are intact. Actually, the mother's influence seems to be even more important than this: the mother tells the embryo not only its head from its rear, but also its belly from its back. A second maternal messenger RNA is introduced into one edge of the unlaid egg which then establishes the top side of the fly, in exactly the same way as *bicoid* dictated the head end.

It seems surprising that the mother fruit fly can exert such tremendous control over her children. These maternal orientation genes certainly do not seem to be compatible with the Mendelian idea of equal parental contribution to offspring. The idea that an embryo's body plan is pre-packaged into the egg echoes the old theory of preformation, which we earlier dismissed so readily. The *bicoid* gene is, of course, hardly a complete system of preformation in the original sense, and clearly a father fruit fly's *bicoid* gene is a fully functional Mendelian gene (it will be required for successful production of his grandchildren), but it is heartening to see echoes of these ancient and apparently discredited ideas springing up again in a modern scientific context.

Maternal control of embryo formation is not just an idiosyncrasy of fruit flies. Within our own vertebrate relatives, it is now thought that mothers can coordinate the early design of their offspring. Zebrafish are a popular model of vertebrate embryonic development, partly because their fry are transparent – we can actually see what is going on. In zebrafish a maternal messenger RNA called *Zf-Sox 19* is pre-packaged into eggs prior to fertilisation and is later found in discrete areas of the developing embryo as early as the eight-cell stage, indicating that its distribution may control the future of these cells.

The large eggs of the *Xenopus* frog are perfect for studying maternal control, because they have an obvious orientation even before they are fertilised, and this orientation is known to affect directly the eventual destiny of cells in the embryo. And here again, we have strong evidence of maternal control as a messenger RNA called *VegT* is inserted into one half of the egg before fertilisation and persists in the embryonic cells derived from that part of the egg. Artificial removal of *VegT* from the embryo has disastrous effects on the embryo's ability to arrange itself: portions of the embryo that should have been brain become skin, portions that should have been gut become connective tissue, and so on. Clearly mother's guiding hand is essential to make a normal baby tadpole.

Drosophila, zebrafish and *Xenopus* are probably the most intensively studied models of embryonic development, and in all three
we have discovered that maternally derived control systems
implanted into the egg before fertilisation play a major role in
arranging the layout of the embryo's body. All these different
examples of nature's 'overbearing mothers' have led scientists to
question whether human mothers control the formation of their
own babies in the same way.

Certainly the potential is there. A wide variety of maternal
messenger RNAs remain in the early mammalian embryo, and
their gradual decay has been tracked up to and beyond the time
when the embryo's genes start to work. Although the embryo
makes messenger RNAs from its own genes, this embryonic
RNA does not make much protein. Instead most protein is made
from RNA left behind by the mother. In fact, this disuse of new
embryonic RNAs seems to be positively encouraged – the early
embryo is able to discriminate between maternal and embryonic
RNAs and preferentially uses the former.

So the maternal hegemony seems complete. Embryonic genes
lie unused while the use of maternal messenger RNA is actively
promoted. This is why the first few cell divisions of a human
embryo are under maternal, not embryonic, control. The early
embryo is simply not autonomous. As if to emphasise the embryo's
impotence, the eventual hand-over of control to the embryo seems
to be at the mother's whim as well: recent studies have shown that
maternal proteins are needed to allow the embryo to start using its
own genes. Not only does the mother dominate the early embryo,
but she even decides when the embryo is ready to go it alone.

Yet, although the mother's guiding hand is important in getting
the mammalian embryo started, she does not help establish the
embryonic body plan like fly, fish and frog mothers do. The
evidence for this is simple. Scientists have known for some time
that they can remove a single cell from embryos containing
between eight and thirty-two cells without adversely affecting
their future development. (This is why the much-feared 'designer
babies' are possible – by studying the genes carried in that one cell,
we can decide whether to transfer the embryo into its mother's

uterus.) By this stage, the period of maternal control is over, but the fact that a cell can be removed without ill effects shows that the cells have not yet been allocated specific destinies. So, even after maternal control is over, embryonic cells still do not know what they are going to become. This means that human mothers probably do not control the fate of particular cells in their embryos.

So mammals are unusual. Mammal mothers do not use their period of control of the embryo to exert undue influence over their children's bodies. Instead, it seems that their interference is aimed solely at shepherding their embryos through their first few hectic days. We do not really know why mammalian mothers do not feel the need to arrange the anatomy of their offspring. Of course, one big difference between us mammals and all those fruit flies, fish and frogs is that our early embryos are safe – they are not stuck on to a piece of rotting banana or drifting around in a pond. Instead, mammalian embryos are cosseted within their mothers, safe from predators and the elements. Because of this, embryonic mammals have different priorities – attaching to their mother and making a placenta are more pressing problems than establishing their body plan. In contrast, flies, fish and frogs must form, hatch and get away as soon as possible, and any help their mothers can give them to speed up this process can only be beneficial.

This first chapter may have come across as rather strange, even unpalatable. The thought of having sex to shed parasites, producing sperm to stop ourselves being overwhelmed by our bacterial tenants, or even comparing fruit flies with our own darling offspring can seem bizarre. At least we have learnt why we have to have sex to have children, and why women and not men get to do all the painful bits. Anyway, our tour of the unusual reproductive habits of the animal kingdom provides an interesting perspective for later chapters.

So where have we reached in our story of pregnancy? Not far in time, I admit, but far in achievement. The mother and father's genes are well and truly mixed and a separate individual has been created. This individual sped through a few rounds of cell division before its maternal momentum ran out. Now, like a car that has stalled in the fast lane of the motorway, it must restart the ignition and forge onwards. The embryo's greatest challenge lies immediately ahead. So far, its mother does not know whether her egg was even fertilised. Now, our tiny embryo must inform her of its presence, or disaster awaits.

○ ○ ○ ○ ○ ○ ○ ○

*How does a mother
'know' she is pregnant?*

○ ○ ○ ○ ○ ○ ○ ○

CHAPTER 2

Breaking the cycle

∘ ∘ ∘ ∘ ∘ ∘ ∘ ∘ ∘

Born on April Fools' Day 1578 in Folkestone, William Harvey was destined to change the face of western medical science. Although he continually struggled with the rudimentary scientific tools available to him, his life was peppered with novel insights and discoveries, and he almost single-handedly laid the foundations of modern medicine. Without recourse to the legends invoked by other natural scientists of the period, Harvey established the basis of modern physiology by the age of fifty and published what was effectively a manifesto for the whole future of reproductive biology by the age of seventy-three.

In this chapter we will be looking at how pregnancy starts – not how an embryo is formed, but how it makes its mother pregnant. Harvey was one of the first people to realise that there are two sides to every pregnancy, and just as much as a pregnancy needs a baby, pregnancy also needs a mother. With hindsight, it is now clear that the musings of this seventeenth-century London physician started a scientific quest that led to our present-day understanding of why women's bodies work the way they do.

Judging by the education bestowed by Thomas and Joan Harvey upon their seven sons and two daughters, the old Kentish Harvey family was a prosperous one. We know little of William's early life until he started his academic career at the King's School in the walled city of Canterbury in 1588. The studious atmosphere of the school was conducive to young Harvey's intellect and in 1593 he took up a place at Gonville and Caius College in the intellectual ferment of Cambridge University.

After Cambridge, Harvey made his crucial move to Padua in Italy. Like a Mediterranean Cambridge, Padua was a university town picturesquely situated on a loop of a river, and it was here that Harvey came to study anatomy and medicine at the turn of the century. Italy is home to the oldest universities in Europe, and Padua had been an academic centre since at least 1222. It had certainly been a centre of anatomical learning for some time by Harvey's arrival: previous leading lights included Andreas Vesalius, writer of the influential early anatomy text *De Corporis Humani Fabrica* (On the Fabric of the Human Body), and Gabriello Fallopio – he of the tubes. Also, the town and its environs were to provide many of the raw materials for Harvey's interest in anatomy. In his later writings, he recalled his dissections of fish bought from Padua market and his observations of the tiny creatures he caught in the nearby Venetian lagoon. It is easy to imagine how the temperate, cosmopolitan, learned atmosphere of Padua must have inspired Harvey, and in many ways he was an English offshoot of the Mediterranean Renaissance.

Yet his time at Padua was not all about soaking up the local atmosphere, and he embarked on a course of formal teaching under the direction of Girolamo Fabrizio da Aquapendente (often Latinised, rather grandly, to Hieronymus Fabricius), himself a student of Fallopio. Fabrizio shared many of Harvey's interests and in particular promoted the idea of autopsy, literally 'seeing for oneself', when it came to studying anatomy. No longer were the classical authors of Greece and Rome to be taken at face value, but instead every claim must be tested by direct anatomical investigation. This idea of gradual replacement of old theories by new ones based on observation was to pervade Harvey's own later work. It was in this spirit that Harvey later explained his attitude to his own work on reproduction:

Fabricius has indeed recounted many wonderful things on the subject of parturition; for my own part, I think there is more to admire and marvel at in conception. It is a matter, in truth, full of obscurity; yet will I venture to put forth a few things – rather though as questions proposed for solution – that I may not appear to subvert other men's opinions only, without bringing forward anything of my own.

As we will see, 'conception' was indeed a matter, in truth, full of obscurity.

For the moment, however, Harvey's interest was drawn towards the field that was to form the subject of his first great work, the study that was to make him famous. The nature of the movements of the heart and blood was not understood at the time when Harvey was at Padua, although great thinkers had already realised that this was an important gap in their knowledge. Temptingly, it also seemed to be a puzzle that might be amenable to existing scientific techniques. Fabrizio had already added to the debate by discovering valves within the great veins of humans and animals, and in his *Anatomia del Cavallo* (Anatomy of the Horse), Carlo Ruini appeared to have clarified the nature of the blood flow between the chambers of the heart. The blood pulsing around Harvey's tiny lagoon creatures must have seemed a golden opportunity. An understanding of the movement of the blood in humans was a prize within sight.

In 1604, Harvey returned to England, where he lived just outside London in the parish of Saint Martin's-extra-Ludgate. His qualifications from Padua allowed him to enter the College of Physicians on 5 October that year. He was subsequently made a fellow in 1607 and two years after that he became Physician at Saint Bartholomew's Hospital. Bart's is an ancient institution and occupies its original site to this day. Instead of the maze-like enclave of medieval lanes in the middle of a modern city centre that we see today, the hospital was then an appendage of the original walled city surrounded on two sides by the Smithfield marshes. Harvey's time at Bart's was divided between his twin interests in the science and the application of medicine and he was made Lumleian lecturer of anatomy and surgery in 1615.

Although we are going to look primarily at how Harvey's studies contributed to our knowledge of the start of pregnancy, it is impossible to ignore his most famous work. The *Exercitatio Anatomica de Motu Cordis et Sanguinis in Animalibus* (Anatomical Essay Regarding the Movement of the Heart and Blood in Animals) is the opus by which all his other writings are measured and its publication in 1628 was the defining moment of his career.

In a few simple chapters he swept away all the old ideas of the function of the heart and blood vessels. The ideas of Galen, taken as a sort of medical gospel for fourteen centuries, were conclusively replaced in a simple series of closely argued expositions, each unusually succinct by the magniloquent standards of the day. When we look at the changes in the circulation at the end of pregnancy in Chapter 5, we will see how important Harvey's discovery was.

After a brief commentary on the weakness of previous theories of the movement of the heart, Harvey quickly explains his own reasons for writing on the subject. He then describes the heart in terms of a pump driving blood around the body. The right side of the heart sends blood through the lungs to the left side of the heart, and this in turn sends it around the body and back to the right side. His argument was finally mixed into one consistent whole in his eighth chapter where he proposes the essentially circular motion of the blood. In the central passage of *De Motu Cordis*, Harvey proclaims the self-evident, almost Copernican, simplicity and cyclicity of his theory and even compares the cycle of the blood with those of the Earth and the heavens:

We have as much right to call this movement of the blood circular as Aristotle had to say that the air and rain emulate the circular movement of the heavenly bodies. The moist earth, he wrote, is warmed by the sun and gives off vapours which condense as they are carried up aloft and in their condensed form fall again as rain and remoisten the earth, so producing successions of fresh life from it.

This analogy is a clear indication of his belief in the importance of his theory: the heart has become like Galileo's sun, sending the blood into an eternal orbit. Like Galileo and Copernicus before him, Harvey's model replaced its predecessors with an idea so simple that it almost had to be true: the heart pumps blood around the body in a circuit.

This idea of the 'circulation' of the blood was to become central to medicine because it allowed the science of physiology to develop. All body organs constantly interact with the blood and this is why Harvey's idea allowed the study of body function to start in earnest. *De Motu Cordis* is a simple document but it has been incredibly important. When we look at his later works on reproduction, we will have to see them in the context of this first book.

Harvey's work on the circulation has a particular story to tell us about its author. Harvey knew there was a gap in his theory. He had shown that the heart pumps blood out into the great arteries, which branch into smaller and smaller arteries, which eventually disappear into the tissues of the body. Harvey could not see what happened next – the next time he could detect the blood was as it drained from the tissues into the veins. We now know that the technology of Harvey's time was inadequate to see how blood traverses the tissues of the body from the arteries to the veins. However, based on the observation that the amount of blood entering a tissue is similar to the amount draining from it, Harvey proposed an as yet undetectable meshwork of tiny vessels that carry out the transfer of blood, and which in doing so provide sustenance to the body. His prediction has turned out to be absolutely correct and today we call these tiny vessels 'capillaries'.

What is impressive is that Harvey was able to overcome the limitations of his techniques and to present his circulatory theory as the best explanation of the unobservable. His argument about the volume of blood flowing into and out of tissues is indirect, but nonetheless clear and compelling. Perhaps most importantly, his theory was inherently simpler than competing theories. Harvey realised that simplicity could be worth more than immediate 'provability' and that was his masterstroke. This great observer was able to distinguish the unobservable from the false and this allowed him to propose such an elegant model. Harvey's ability to juggle simplicity and observability was to be even more important in his work on pregnancy – so much more of this biological story was impossible to demonstrate using seventeenth-century techniques.

Harvey: the generation game

William Harvey retired from his formal duties in 1646, but it seems that his drive was undiminished and his discoveries continued. Of his two further books, one, a study of insects, was lost during the English Civil War. The other, his treatise on reproduction, appeared in 1651 with the simultaneous publication in Cambridge and Amsterdam of the *Exercitationes de Generatione Animalium, de Partu, de Membranis ac Humoribus Uteri et de Conceptione* (Essays Regarding the Generation of Animals; On Birth, the Uterine Membranes and Humours and Conception). In it, Harvey addressed the fundamental questions of animal reproduction. How is a new life made? What is the contribution of the mother and father? How does an embryo persuade its mother to nurture it? Three and a half centuries later, these questions remain as important as they were in Harvey's time, and none of them has been fully answered.

In confronting the mysteries of reproduction, Harvey was now embarking on a scientific odyssey that he could never complete, and it seems that he knew it. Instead of the neat, compact, almost smug chapters of *De Motu Cordis*, *De Generatione* is an altogether more expansive affair. The irregular structure of the book is almost an admission of the enormity of the challenge facing Harvey. Often he seems to be using *De Generatione* to raise questions rather than to provide answers. For those who feel that the intellectual chase is more fun than the kill, this is what makes *De Generatione* Harvey's most inspiring work. Modern readers know that Harvey cannot answer his own questions, or even observe the processes he wishes to observe, but his continual struggle against these problems makes his book on reproduction his most compelling.

First, Harvey tackled the nature of generation itself. In the seventeenth century, no one knew where new animals spring from. Of course, we now know that babies are produced when males' sperm and females' eggs combine to form an embryo. Before the advent of the microscope, however, this process was invisible and scientists were left to speculate on the miraculous appearance of offspring in the womb. With hindsight, we can see that Harvey had two main problems: he could not see the sperm cells in semen and he could not see mammalian eggs. He was clearly confused by the apparently formless nature of semen and the lack of a mammalian equivalent to the large eggs of birds.

Another problem was that 'spontaneous generation', the idea that life can spring forth from nonliving matter, was a widely held belief in Harvey's time. Flies seemed to arise directly from filth, mosquitoes from stagnant water and even, according to one story, geese from barnacles on the keels of boats (hence the barnacle goose). Not fully refuted until the experiments of Louis Pasteur in the nineteenth century, spontaneous generation clouded scientists' thinking even when they studied larger animals. It was obvious that pregnancy followed on in some way from sexual intercourse in people and in farm animals, but it was by no means certain that the contributions of the parents were physical rather than spiritual.

Harvey's difficulty in demonstrating physical products of parents which fuse to produce offspring plagues *De Generatione*. It did not help matters that half of the book describes reproduction in birds, in which females produce an obvious physical entity, the egg, which is her contribution to the chick. The bird's egg fitted Harvey's idea of physical parental contribution to offspring, but the apparent lack of such an entity in mammals was a big problem for him.

Despite the practical limitations facing Harvey in his quest, he realised that physical contribution by the parents was still a distinct possibility. Semen was an obvious candidate for fathers' contribution to babies, but Harvey had no microscope with which to study it. He spent considerable time discussing the nature of semen, but he seems to have remained undecided on its role in reproduction. This may have been partly because he could not find an equivalent fluid made by females – to Harvey, the fact that females do not make semen seemed to cast doubt on the importance of the semen of the male.

Left with the problem of no 'female semen' and no obvious mechanism by which male semen can act, Harvey realised that the theory that parents make a physical contribution to their offspring was in trouble. Try as he might, as much as he believed that maternal and paternal material gave rise to the embryo, he could not find that material. Nonetheless, Harvey was aware that this did not mean that this material did not exist, but that he simply might not be able to observe it. Just as he was able to accept the existence of invisible capillaries and proceed to a complete understanding of the circulation of the blood, he must have felt that the invisibility of the male and female germinal material was not an insuperable problem.

Harvey and the start of pregnancy

Harvey emerged from his failed attempt to explain the origin of embryos unscathed. In his own mind, he could set aside the unanswered question of the role of male and female semen and continue with his investigation of pregnancy. As he reaches the end of his discussion of the origins of the embryo, he hints at another unanswered question – a question that remains important to this day, and one that he was to state and investigate more clearly than anyone else: 'the woman, after contact with the spermatic fluid *in coitu,* seems to receive influence, and to become fecundated without the co-operation of any sensible corporeal agent, in the same way as iron touched by the magnet is endowed with its powers and can attract other iron to itself'.

With this magnet analogy, Harvey is suggesting that mothers and embryos are equal partners in pregnancy, and that the mother must not be ignored in scientists' desperate search for the embryo. He is claiming that the start of pregnancy may be as much a change in a woman as the formation of a child. Before an embryo can be fully conceived, does a woman have to be 'made pregnant' in some way? Between his time and ours, we have discovered that changes wrought in the mother to 'make her pregnant' are indeed just as important as the formation of the embryo. After all, without a cooperative host the baby is doomed. Harvey recognised that pregnancy is a partnership, so he dedicated a great deal of effort to studying how mothers are persuaded to carry a child.

In his relentlessly logical way, Harvey began to break the problem of the start of pregnancy down into its constituent parts.

First of all, he simplified matters considerably by realising that different animals breed in fundamentally different ways, and that female animals that lay eggs may never need to be 'made pregnant' in the way he was suggesting. He knew that birds can lay eggs whether they have been mated or not, so presumably they never need to 'know' if they are pregnant at all. Because of this, he started to direct his attention towards animals that bear live young – mostly mammals – and tried to discover how they are forewarned of impending pregnancy.

Harvey's intellectual dismantling of conception now needed a convenient animal on which to work, and the animal he chose was the deer. In 1630 he became personal physician to Charles I, and this seems to have involved accompanying the monarch on his hunting trips, allowing him access to large numbers of freshly killed deer. Deer were also a good choice because a great deal was known about their sexual behaviour.

Harvey was clearly impressed by a particular feature of deer reproduction that he thought might help him in his quest: all female deer start to breed at the same time of year, in the autumn. Although other domestic animals, such as horses and sheep, breed better at certain times of the year, the seasonality of the love life of deer is particularly dramatic. Today, if we want to study breeding in a group of female mammals, we usually try to bring them all into heat at the same time. This allows them to be compared one to another, and followed over time as a synchronised group. By studying deer, Harvey had access to large groups of females that were already naturally synchronised. If you study deer, nature does the first part of your experiments for you.

In the strictly seasonal deer, the relationship of intercourse to birth is obvious, because female deer all mate around the same time and all give birth around the same time, in the summer. Not only do female deer obligingly come into heat together, but the time of mating can be pinpointed exactly because deer sex is a very quick affair. Because of this Harvey was able to calculate the length of pregnancy in red deer to be 231 days. He then confirmed in deer what he had found in other species: that he was unable to find any embryo in the uterus early on in pregnancy. Harvey looked long and hard to find a physical product of conception in deer, a mammalian equivalent of the bird's 'egg', but he repeatedly failed to find one for weeks after mating:

Nor is that true which Aristotle often affirms, and which physicians take for granted, namely, that immediately after intercourse, something either of the foetus or the conception may be found in the uterus.

In the way above indicated do the hind and doe, affected by a kind of contagion, finally conceive and produce primordia, of the nature of eggs, or the seed of plants, or the fruit of trees, although for a whole month they had exhibited nothing within the uterus.

Nothing . . . can be discovered there – neither the semen of the male, nor aught else having reference to the conception – during the whole of the months of September and October.

He was adamant on this point: he could first find evidence of the embryo some time after the deer hind had failed to come back into heat. He immediately started to question what had been going on between mating and this belated appearance of the embryo. How can a deer hind be pregnant without an embryo?

Harvey now tried to relate his findings to humans. Failure to come back into heat is the outward sign that deer are pregnant, but women do not show periods of heat. Unlike most female mammals, women are receptive to the opposite sex throughout their cycle. Instead, pregnancy in women is heralded by a cessation of menstruation. Despite this fundamental difference, Harvey could see that both women and deer hinds 'know' very quickly when they are pregnant: women do not menstruate after conceiving and pregnant deer fail to come into heat after a fertile mating. But how do mothers 'know' so soon that they are pregnant? This great question was Harvey's legacy to reproductive biology.

Harvey's lack of a microscope can almost be seen as fortuitous because, if he had been able to see that the embryo is, in fact, present almost immediately after intercourse, he may never have delved so deeply into the maternal side of pregnancy. We now know that, even though an embryo is formed within a day or two of copulation, its first job is indeed to inform its mother of its presence.

To Harvey, the importance of the maternal side of conception must have seemed overwhelming. First of all, Harvey made it clear that the impregnation of a woman might involve changes in her entire body, and not just her womb. As we will see later, this suggestion has proved to be uncannily accurate:

To what is this active power of the male committed? is it to the uterus solely, or to the whole woman? or is it to the uterus primarily and to the woman secondarily? or, lastly, does the woman conceive in the womb, as we see by the eye and think by the brain?

Incidentally, the above quotation is by no means Harvey's only comparison of the uterus with the brain. It is no coincidence that both organs are said to 'conceive', and he explains why this is so:

and since the structure of the uterus, when ready to conceive, is very like the structure of the brain, why should we not suppose that the function of both is similar, and that there is excited by coitus within the uterus a something identical with, or at least analogous to, an imagination or a desire in the brain, whence comes the generation or procreation of the ovum? For the functions of both are termed 'conceptions'.

That both the mind and the womb are said to 'conceive' may now seem to be just a linguistic quirk, but why should the conceptions of the two organs not be similar? Certainly, both processes were invisible to Harvey. Importantly, this suggestion of a 'mental' nature of uterine conception seemed to encourage him to question how the uterus becomes 'aware' of pregnancy.

Unfortunately, at this point Harvey's powers of observation seem to have let him down. We now know that the structure which orchestrates the start of pregnancy is continually visible on the surface of the ovary from conception to birth. Rather surprisingly for someone who spent so much time dissecting female mammals, Harvey made a grave error in asserting that he could see nothing of interest on the ovaries. He even seems to doubt that these 'female testicles' have anything to do with procreation at all:

> I, for my part, greatly wonder how any one can believe that from parts so imperfect and obscure, a fluid like the semen, so elaborate, concoct and vivifying, can ever be produced, endowed with force and spirit and generative influence adequate to overcome that of the male . . . the female testicles, as they are called, whether they be examined before or after intercourse, neither swell nor vary from their usual condition; they show no trace of being the slightest use either in the business of intercourse or in that of generation.

While Harvey can be excused for failing to find semen and embryos in the uterus, it is a mystery why he failed to notice the ferment going on in deer ovaries during the breeding season. Just this one observation and Harvey could have solved the pregnancy problem in its entirety. But it was not to be.

However, Harvey's basic question has come down through the centuries to modern reproductive biologists almost unchanged: how does the embryo announce its presence to its mother? Without this announcement, the embryo will be destroyed. The problem arises because, unlike males, the mammalian female's reproductive organs have a double function – they act as an arena for two different processes: fertilisation and pregnancy. In most mammals these processes cannot go on simultaneously, so the embryo must arrest its mother's attempts at fertilisation so that it can start a pregnancy. It is this double role of the female reproductive tract that the mammalian embryo has to control, because it has to switch its mother from a cycling sexual animal to a more stable nurturing one.

This is where we left our embryo at the end of Chapter 1: genetically autonomous, but with a tremendous challenge ahead before it can start to develop. The challenge is to arrest its mother's sexual cycles and prevent the next period of heat (deer) or menstruation (humans). For conception to be successful, something other than just fertilisation must occur: the mother must recognise her offspring's presence and prepare to accept it. So the pressure is on for the heroic embryo. Millions of times smaller than its host, it must completely divert her body processes within a few weeks. How it achieves this daunting task is in many ways the most impressive part of pregnancy.

Harvey did not realise that the embryo was even present at the time of this embryonic take-over, and so he made the most sensible guess based on the available evidence. He suggested that females were converted from the sexual to nurturing state by some force imparted by semen to their entire body. As we shall see, in some mammals (not us or deer, though) female sexual activity is brought to a halt merely by the act of copulation. As we will also see, however, embryos have to take the lead in starting pregnancy in women and deer hinds. These embryos are the drivers, not the passengers.

Reproduction after Harvey

Harvey died just before the development of the microscope, and this has left him suspended in a kind of historical limbo – he was not to see his promised land. This new instrument led to a spate of dramatic advances in biology. In 1677 the nature of semen was among the first biological questions to be answered by the microscope when its inventor, Anton van Leeuwenhoek, first described the tiny tadpole-like creatures that populate seminal fluid. Yet the discovery of sperm was not universally accepted as evidence that they were involved in reproduction – one theory, which was popular for the next 200 years, stated that sperm were actually a type of parasite that specialised in infesting the testicle.

The 'seminal' matter of the female was discovered at about the same time. In his *De Mulierum Organis Generationis Inservientibus* (Regarding the Generative Organs of Women) of 1672, Reinier de Graaf first correctly ascribed a reproductive function to the blister-like structures that pepper the ovaries, and which are now called Graafian follicles. Eventually, these follicles were found to produce that elusive creature, the human egg itself, by Karl Ernst von Baer in 1827 and described in his *De Ovo Mammalium et Hominis Genesi* (On the Mammalian Egg and the Genesis of Man). At last it was now understood how a female generative particle, an egg, is ejected from a woman's ovaries every month. From this time, with Harvey's script and these microscopic *dramatis personae*, the drama of reproductive biology was able to start in earnest.

As our story of reproductive biology nears the twentieth century, we approach the crucial discovery that finally did away with all the ephemeral humours, influences and spirits that litter the old texts on reproduction. Early in the century these were replaced by hormones – the chemical messengers that allow all the reproductive organs to communicate with each other. Hormones are central to our story because they are the way in which the reproductive system is bound together into a cohesive, efficient machine.

Hormones are secreted into the bloodstream by internal or endocrine glands, and they exert their effects on other, distant organs. Hormones were first discovered when scientists ground up the endocrine glands from one animal and injected them into another to see what happened. Different glands seemed to release one or more chemicals, each with distinctive effects. There were hormones that made goats squirt milk from their udders, hormones that made sheep behave as if they were on heat, and hormones that ended pregnancy. The whole jigsaw of reproduction must have seemed to be falling into place before the early endocrinologists' eyes. They spent much of their time purifying hormones from animal tissues and trying to work out their effects. It soon became clear that the female reproductive system, whether cycling or pregnant, was controlled by just a few crucial hormones. This remarkable fact means that by measuring the blood levels of just a few chemicals, we can watch the machinery of female reproduction in action.

When scientists studied the chemistry of the hormones that control female reproduction, they turned out to be a rather heterogeneous bunch. Some are oily chemicals derived from cholesterol, and these are now called steroids. Most other hormones are proteins of a variety of shapes and sizes. Although there seems to be no rhyme or reason to the chemistry of female reproductive hormones, the female body is obviously expert at telling them apart and this is why hormones can exert such specific effects. For example, a hormone poured into the bloodstream by the ovary will be pumped around the body until it is picked up by an organ that bears the appropriate 'receptor' molecule for that hormone. That organ then changes its behaviour in a way dictated by the hormonal signal it receives. Hence, by careful arrangement of hormone-producing and hormone-sensitive cells around the body, tissues that are not in physical contact with each other are able to work in concert. This hormonal internet allows all our organs to be in constant communication.

A reproductive revolution started in the 1960s when scientists developed an ingenious and extremely sensitive way of measuring amounts of a single hormone in small blood samples using radio-activity. This 'radioimmunoassay' technique has proved extremely useful because it means we can now follow the patterns of hormone production over time to find out how different hormones interact. Radioimmunoassay is quite quick and so one person can do hundreds of assays at a time (believe me, I know). Also the technique is so sensitive that tiny quantities of hormone can be detected, so you do not need to take an enormous amount of blood from an animal to find out what its hormones are doing. Because of their sensitivity and speed, much of what we know about reproduction is based on radioimmunoassays.

By the end of the 1960s, scientists had identified most of the building blocks of the female reproductive system: they knew about eggs, sperm, embryos and hormones, and they also had a good way of measuring the changes in hormones in healthy animals and people. Now they could observe the machinery controlling the female cycle, and with this knowledge start to discover how the embryo stops that cycle.

The ups and downs of the cycle

The reproductive cycles of women and deer are controlled by a hormonal 'conversation' between several different organs. We will start with a crop of Graafian follicles ripening on the ovaries, each with their valuable cargo of a single egg.

As well as containing an egg, each follicle is also lined with a thick layer of cells that secrete large amounts of hormones. The best-known follicular hormones is the estrogen – a group of steroid hormones that help to control the female body. Estrogen got its name because it induces sexual activity, or estrus, in female sheep. (Estrus behaviour is in turn named after a fly, *Oestrus ovis*, that annoys sheep. Apparently the frisky behaviour of ewes worried by the fly is similar to their behaviour when they want to attract rams.)

Estrogen is in many ways the 'feminising' hormone. It is instrumental in inducing the development and growth of the uterus, vagina and breasts in girls, and helps to control these organs throughout pre-menopausal life. Estrogen also promotes the development of female characteristics not directly related to the reproductive organs, such as a wider pelvis and the pattern of fat distribution typical of women. Like most steroids, the effects of estrogen is all-pervasive and the female body is constantly awash with fluctuating levels of this highly potent hormone.

The amount of estrogen secreted by the Graafian follicles on the ovaries varies dramatically during the cycle, and this variation is controlled by the brain. It is thought that, as vertebrates have evolved, their brains have become more and more involved in reproduction to allow them to attune their breeding behaviour to external factors, such as seasonal, nutritional and social changes. For example, connecting the reproductive system to the brain means that females can take factors such as the time of year or the presence of other offspring into account when they make their reproductive 'decisions'. Mammals are no exception to this, and their ovarian activity is kept on a tight rein by the all-controlling pituitary gland dangling from the underside of the brain. The pituitary is an endocrine gland that interprets information from the brain and disseminates it in the form of hormonal signals to the rest of the body (Figure 1). The pituitary is the reproductive organs' window on the world.

Figure 1. The hormones that control female reproduction. The organs and hormones connected by thick lines are involved in controlling the menstrual cycle and pregnancy (discussed in Chapter 2) and those connected by thin lines control labour and lactation (discussed in Chapter 5).

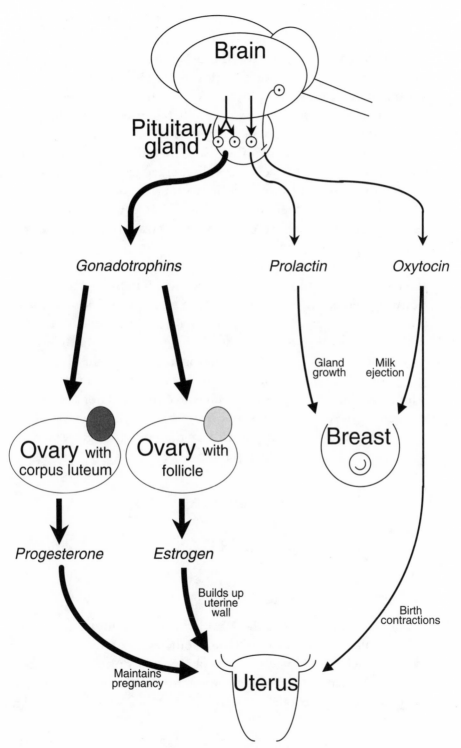

Two of these pituitary hormones (called the follicle-stimulating hormone and the luteinizing hormone) drive ovarian activity during the cycle. These two hormones are sugary proteins and are also called gonadotrophins ('ovary/testicle nourishers'). Between them, these two pituitary gonadotrophin hormones can stimulate ovarian follicles to grow all the way up to the point where they burst, or ovulate, releasing an egg. Because the gonadotrophins are the main way in which the brain controls the ovaries, their release is very carefully controlled. Instead of being continuously poured into the circulation, they are released in a staccato fashion, at the rate of several pulses per day. This pulsed release is driven by a hormone produced by a group of nerve cells in the lower part of the brain.

An important feature of gonadotrophin secretion is that the ticking gonadotrophin clock does not just blindly dictate ovarian hormone secretion – it also responds to what the ovaries are doing. There is a continual two-way hormonal conversation between the brain and the ovaries that controls the activity of both organs. Gonadotrophins cause ovarian estrogen release, but this estrogen then acts on the pituitary to suppress gonadotrophin secretion. This balancing act ensures that secretion of the two hormones does not get out of control. This simple form of self-regulation is known as negative feedback, and it is used by the body to regulate a wide variety of processes in which stability is required.

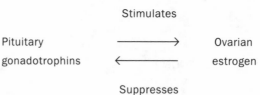

So, gonadotrophins from the pituitary drive the initial growth of ovarian follicles. As they grow, these follicles start to produce more estrogen, which then starts to suppress gonadotropin secretion by the pituitary. This suppression ensures that growth of the follicles does not get out of control. As the follicles keep growing, however, they gradually produce more and more estrogen, so that the concentrations of estrogen in the blood rise.

When blood estrogen reaches a critical concentration, something rather unexpected happens: the negative feedback control breaks down. As estrogen secretion passes this threshold level, it suddenly starts to stimulate, rather than suppress, the release of pituitary gonadotrophins. Now estrogen is stimulating gonadotrophin secretion and gonadotrophins are stimulating release of estrogen, and the secretion of both shoots up in an almost uncontrolled way.

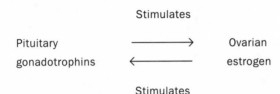

This change to escalating positive feedback is a central feature of female reproduction, because the resulting surge in gonadotrophins triggers the final maturation of the egg and causes ovulation. In some non-human species, the surge of estrogen also acts on the brain to induce a period of 'heat' in females. Heat or estrus is a period during the sexual cycle of females at which they will accept sexual advances from males. Estrogen is a good trigger to use for sexual activity, because it is released in the greatest quantity just before the follicle ovulates and releases the egg. Because of this, the female will mate at a time when an egg is available for fertilisation.

Estrogen does not induce sexual behaviour in women, and this is probably because women do not have a discrete period of heat during their cycle. Instead women are unusual in that they often have sex throughout the cycle, although we are not sure why. Most other female mammals seem to know when they are fertile, and it seems rather inefficient for women not to know. Instead of mating at the time when a freshly ovulated egg is available, they just mate at random. One possible reason is that women do not know when they are fertile so that their partners will not know either – perhaps this uncertainty helps to keep men close to their partners, meaning that they will be more likely to help bring up the children. At the other end of the marital-faithfulness spectrum, another theory claims that the lack of a definite heat period means that a promiscuous woman can mate with many men and none of them will know if the resulting child is theirs – and thus will have no desire ever to harm it, because they will not know if they are its father. For whatever reasons women lost their ability to know when they are fertile, the change seems to have been based on the assumption that men are a pretty disreputable lot.

By dint of an elegant dance of hormonal interactions, the ovary and brain have guided a follicle all the way to maturity and ovulation. Ovulation is the climax of the cycle, and its *raison d'être*. Despite its importance, the actual processes involved in ovulation are not completely known. There is a build–up of pressure in the follicle, but this is probably not the actual cause of follicular rupture. Probably, the wall of the follicle is weakened in some way. Whatever the cause of ovulation, once the egg is free of the follicle it is drawn into the Fallopian tubes and off to its fate.

The ruptured follicle remains behind on the ovary, but its contribution to pregnancy is far from over. Before ovulation, the female's hormonal changes were all aimed at releasing a healthy egg into an environment conducive to its fertilisation. Now, after ovulation, the ruptured follicle also has a vital role in preparing the uterus to receive the embryo. Ovulation is the instant when the uterus changes from being a site for fertilisation to a site for pregnancy.

As the follicle collapses after ovulation, its hormone–secreting cells start to grow. The hollow follicle fills in and fattens up and starts to look slightly yellow in some species, becoming the corpus luteum ('yellow body'). The gonadotrophin surge has wrought great changes in the chemistry of the cells of the corpus luteum, and instead of making estrogen, they now start to produce progesterone ('for pregnancy'). Like estrogen, progesterone is a steroid hormone, and has many different effects throughout the body. Made as a direct result of ovulation, the corpus luteum is a central player in our pregnancy story.

During the reproductive cycle of both humans and deer, the corpus luteum secretes progesterone for a prolonged period – the corpus makes progesterone for nearly half the cycle in humans and for perhaps 80 per cent of the cycle in deer. Like estrogen, the secretion of progesterone is stimulated by gonadotrophins from the pituitary, and the system is again regulated by negative feedback, because progesterone suppresses gonadotrophin secretion.

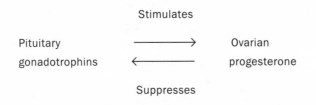

Unlike estrogen, however, the regulatory control of progesterone never fails in the same way as it did for estrogen. Because of this stability, the corpus luteum secretes progesterone steadily throughout much of the cycle.

This luteal progesterone has two main jobs, both of which are essential for a successful pregnancy. First, by keeping gonadotrophin secretion in check, the corpus luteum ensures that no fresh ovarian follicles reach maturity. A new ovulation would cause tremendous hormonal disruption to the embryo and it would almost certainly be lost. Second, progesterone also has a positive contribution to make to the embryo's future. Progesterone helps to convert the uterus to a 'nurturing' state, by changing its structure and chemistry. These changes are most dramatic in the great apes (including us), in which the inside lining of the uterus builds up into a thick glandular layer to receive the embryo. This tissue is established under the control of estrogen, but luteal progesterone then induces it to secrete nutrients for the embryo.

Of course, more often than not, women do not become pregnant during a cycle. If there is no fertilisation, and no embryo, then the female has to ovulate again. A new ovulation can take place only if the old progesterone-secreting corpus luteum is destroyed – a process known as luteolysis. The corpus luteum has a fixed life span: an inbuilt obsolescence that allows the cycle to restart once the corpus luteum has served its purpose. As we will see, the destruction of the corpus luteum is a turning point for a female animal – a time when she either becomes pregnant or has to start trying to conceive again.

In many animals, including deer, what happens after luteolysis is fairly straightforward. The concentration of progesterone in the blood plummets, and progesterone stops suppressing pituitary gonadotrophin secretion. As gonadotrophin levels in the blood start to rise, ovarian follicular development can start once more. Now we are back where we started with a crop of developing ovarian follicles: no baby last time, so it is time to try again.

So this is what happens during female reproductive cycles in most animals. The cycle grinds on indefinitely until it is switched off by pregnancy, senility, emaciation or the end of the breeding season. As long as the individual components of the system (a fresh supply of follicles and a supportive brain) remain intact, the female reproductive cycle keeps on turning as surely as night follows day.

Why women are different

The inexorable progression of women's reproductive cycles clearly impressed the ancients and they ascribed almost religious significance to the cycle of fertility. Not only is the human menstrual cycle set in a religious context in the Bible, but the correspondence of the cycle to the time taken for the Moon to travel around the heavens has been enshrined in the word 'menstrual' ('monthly').

Menstruation itself, however, has rather a bad press from mystics and philosophers who were, and still are, predominantly male. The fact that female animals do not menstruate obviously suggested to many men that women are somehow unusual, even slightly suspect. Judaism and Islam both hold menstruation to be unclean – the Old Testament lays down strict rules about women keeping away from men during menstruation, and the Koran describes menstruation as 'pollution'. Pliny asserted that menstrual blood causes bees to leave their hives, mares to miscarry, fruit to fall from trees, the edge of a razor to dull, wine to go sour and grass to wither. Surely there is material here for tampon advertisements.

The Greeks had rather more respect for menstruation. Many followers of Aristotle believed that the menstrual blood was a woman's contribution to pregnancy. The semen was said to descend on it, rather like a seed falls on fertile ground, and from this union a child was made. The evidence to support this idea was that women who fail to become pregnant discard their menstrual blood, whereas those who become pregnant 'retain' it.

These ancient musings highlight an important distinction between women and domestic animals that has been obvious but unexplained for millennia. Why do women menstruate? As it turns out, menstruation is not reserved solely for women, but is a characteristic of female 'old world' primates in general. The menstruation question is important because it is inextricably linked with the nature of human pregnancy.

We think we now know how menstruation happens, although of course this does not explain why. Menstruation takes place just after the demise of the corpus luteum, as progesterone levels start to fall. The thick, blood-engorged lining of the uterus had been maintained mainly by progesterone, so when this hormone disappears after luteolysis, the lining falls off, causing the menstrual flow. After the start of menstruation, women do not bounce back to ovulate within a few days in the way that most mammals do after luteolysis. Instead, the delay between menstruation and ovulation is around two weeks. (As the lifespan of the corpus luteum that preceded this period was also about two weeks, ovulation and menstruation are usually set apart by about two weeks within a four-week cycle.)

We are not certain why old world primates evolved menstruation. I am sure many women readers will be hoping that there is a good reason for it, but I am afraid the jury is still out on this one. One suggestion is that menstruation has a cleansing effect to rid the uterus of germs introduced by the male during intercourse. Why only a few species of monkey should be in need of these hygiene precautions is unclear. Menstruation does not seem to be more copious in more promiscuous primate species and the finding that the bacterial load in the uterus is not greatly reduced after menstruation suggests that menstruation is not a hygiene precaution at all.

A theory that menstruation saves energy seems more convincing. The fully grown uterine lining is a profligate user of precious energy, and this has led to the idea that it is shed at menstruation to save energy. Women's metabolism runs faster when the uterine wall is fully developed and it has been calculated that by discarding it for half of her cycle, a woman saves around one-and-a-half days' food. Unfortunately, the energy-saving theory raises yet another difficult question about menstruation: if resources are so valuable, then why do women not just absorb their uterine lining and reuse it for their next cycle, rather than throwing it away?

More recently, it has been proposed that menstruation is not actually 'useful' at all, but has instead evolved as a side-effect of the way human pregnancy works. The human embryo is unusually aggressive in the way it invades the wall of the uterus, and it has been suggested that women have evolved an extra-thick uterine lining to cope with this aggressive invader. This 'maternal buffer-zone' idea has attracted much support among pregnancy biologists because it links two inexplicably unusual human characteristics: embryonic invasiveness and menstruation. Maybe the human uterine lining has become so thick that it can no longer be reabsorbed and instead must be shed. This would mean that menstruation developed not for any particular function, but as an easy way out of the problem of disposing of a thickened uterine lining.

For some women, menstruation is much more than just an unpleasant inconvenience – it can make their life a continual struggle with pain, weakness and infertility. Between 7 and 15 per cent of women are thought to be affected by endometriosis, making it one of the commonest diseases of mankind. In endometriosis, the thickened lining of the uterus becomes almost as aggressive as the fetus it was designed to absorb, and fragments of uterine lining spread around the abdomen, attacking the Fallopian tubes, the ovary, the intestine and the bladder, and even infesting surgical scars. Pockets of tissue can also colonise the lungs, skin, limbs or even the brain, as the uterine lining starts to behave almost like a tumour. The exceptional ability of the human womb lining to proliferate is now turned against the rest of the woman's body, as monthly hormone cycles drive endless rounds of futile menstruation and haemorrhage in scattered colonies of uterine cells.

When normal menstruation takes place in the uterus, the discarded blood and debris are lost though the vagina, but when it happens in the abdomen, chest, limbs or head, there is nowhere for the waste material to go. Wherever it appears, the misplaced menstrual flow irritates nearby tissues, causing pain, inflammation and scarring. The commonest symptoms of this bizarre colonisation of a woman's body are painful menstruation, pain during sex and a feeling of weakness. Also, she may show signs of damage to colonised organs: constipation, diarrhea, painful urination, stiffness and even coughing up blood. Endometriosis is also a major cause of infertility and perhaps 30 or 40 per cent of sufferers have blocked Fallopian tubes. Even if they become pregnant, women with endometriosis are more likely to have a miscarriage.

It seems that humans have made a terrible miscalculation in producing such an exuberant uterine lining for their babies, but still no one knows why endometriosis happens. The most popular theory is that uterine cells shed at menstruation sometimes leak up the Fallopian tubes instead of passing into the vagina. From the tubes they escape into the abdomen and 'seed' the first organ that they meet. It has been suggested, however, that this retrograde menstruation is actually a normal occurrence, and that the cells that reach the abdomen simply die in most women. Possibly, women with endometriosis lack some hormonal or immune mechanism to destroy these cells.

Yet, it is difficult to see how cells that escape through the Fallopian tubes could end up in the lungs or arms, and it has also been claimed that endometriosis cells might spread from the uterus in the blood, just like many malignant tumours. Another possibility is that endometriosis cells are a kind of 'throwback' to the rapidly dividing cells that originally formed the woman's body when she was an embryo. Perhaps scattered groups of misguided fetal cells decide that they are going to make an island of uterus in a sea of brain, but only spring to life at puberty when they are stimulated by hormones from the ovary.

More recently, suspicions have arisen that endometriosis may be caused by environmental pollutants, such as dioxins and poly-chlorinated biphenyls (PCBs). Of course, it would be a terrible indictment of our cavalier attitude to our industrial poisons if they were a cause of endometriosis, but it would at least explain the disease. In a strange sort of way, an external cause of endometriosis would almost be reassuring – we would no longer have to believe that we had evolved a method of reproduction that inevitably led to something like endometriosis. Is this 'other cancer' really a natural process?

There is no cure for endometriosis – it is difficult to cure something that is so mysterious. Because the growth of endome-triosis colonies is driven by the hormones of the menstrual cycle, however, women can often get some temporary relief by jamming the cycle with various types of contraceptive pill, or even by becoming pregnant. Yet for many the best option is surgery to remove as many of the colonies as possible, although even hysterectomy does not cure the disease if there are colonies on organs other than the uterus. Many sufferers simply have to try their luck with painkillers.

Women have one thing for which they can thank menstruation: it gives them one definite advantage over men. Hereditary haemachromatosis is one of the commonest genetic diseases of humans and it is inherited as a Mendelian recessive trait (that is, sufferers inherit one defective gene from each parent). People with this disease have a defect in iron metabolism which means that they accumulate abnormally high levels of iron in their bodies, with potentially fatal results. Although iron is essential for making blood haemoglobin, it can also be extremely poisonous – sometimes even causing liver failure. The toxicity of iron is what causes the damage in haemachromatosis, but fortunately there is a convenient treatment for the disease – regular bleeding disposes of enough iron to keep the body's accumulation below a dangerous level. Of course, regular bleeding is what women of reproductive age do naturally, and this is why haemachromatosis is less dangerous to women. Women with haemachromatosis treat their disease every month by menstruating. Every cloud has a silver lining.

What the embryo has to do now

After ovulation, a human embryo has just two weeks to stop its mother menstruating. Life is only slightly less hectic for a red deer calf, with perhaps sixteen days to prevent luteolysis.

Obviously, the embryo has everything to gain or lose from this process, but starting a pregnancy is a monumental decision for its mother as well. Is she, or is she not, pregnant? A wrong decision would lead to a huge waste of her resources, even to the extent of jeopardising her chances of ever producing any children in the future. Repeated failure to detect the presence of her tiny offspring could end her chances of passing on any genes at all. From the mother's point of view, each time this decision is made is a turning point in her life.

As we will see in the rest of this chapter, the biological mechanisms involved in this 'maternal recognition of pregnancy' turn out to be quite different in different mammals, presumably reflecting their differing needs. Yet there is one central thread to this process common to all mammals that we have studied so far. The unifying feature of mammalian reproduction is that for pregnancy to start successfully, one or more progesterone-secreting corpora lutea must persist on the ovaries. Progesterone will then continue to suppress further ovulations or menstruations throughout pregnancy, giving the embryo (or embryos) a chance to grow. This is why the corpus luteum is essential for pregnancy, and why many mammalian embryos devote their attention to ensuring its survival.

The corpus luteum is not just important, it also seems to be very ancient. Even in vertebrates that do not have long pregnancies, luteal progesterone still encourages embryonic development and reduces the mother's tendency to expel eggs and embryos. Except in birds, in which luteal progesterone has been side-tracked into having a role in parental behaviour, progesterone helps to promote internal retention and nurturing of embryos. This process has been developed to its zenith in those ultimate nurturers, the mammals.

So, to give their offspring a chance to arrest the sexual cycle, mammalian mothers allow their cycles to be controlled at three points. These three crucial moments of decision are: ovulation, formation of the progesterone-secreting corpus luteum and destruction of the corpus luteum. These three processes are controlled in different ways in different mammalian species. All three can happen spontaneously or they can be controlled by the mother, the father or the embryo. In fact, almost all the amazing diversity of mammalian reproduction is based on variations on this simple three-part theme. Each species has honed the intricacies of ovulation, luteal formation and luteal destruction to fit its own requirements and maximise its chances of success.

So how does the embryo pull off the trick of arresting its mother's cycles? Although different mammals have tackled the problem in different ways, their strategies can be classified into four basic systems, typified by dogs, cats, deer and humans. In some of these animals the parents control the onset of pregnancy and in others the embryo is in charge. Humans are at the 'embryo-in-charge' end of the spectrum, but we will start at the other extreme, with our furry friends.

Strategy 1: some animals assume they are pregnant

The first solution to the problem of ensuring that pregnancy gets under way seems almost too simple. This least human-like system is very straightforward, although, as we will see, it has hidden depths.

William Harvey was the first person to point out that not all animals seem to 'know' they are pregnant. He pointed out that bitches seem to think they are pregnant, even when they are not:

Over-fed bitches, which admit the dog without fecundation following, are nevertheless observed to be sluggish about the time that they should have whelped, and to bark as they do when their time is at hand, also to steal away the whelps from another bitch, to tend and lick them, and also to fight fiercely for them.

Many owners of unneutered bitches will recognise the phenomenon Harvey describes as 'pseudopregnancy'. Yes, dogs are different. Barren bitches will nest in anticipation of puppies that never appear and will even produce milk for them. Their brain seems hard-wired for maternal care and they will nurse other bitches' puppies as well as toys, kittens and whatever else they can lay their paws on. Harvey had succeeded in dividing mammals' pregnancy strategies into two fundamental groups: those that know when they are pregnant, and those that just assume.

Women can suffer from a condition called pseudopregnancy, but it is quite different. Whereas pseudopregnancy is a common and natural condition in dogs, human pseudopregnancy is at root a psychiatric disorder. It appears that a strong desire to get pregnant, a fear of losing a partner or even low self-esteem can cause the brain and pituitary to release a barrage of hormones that mimic pregnancy. Women with this problem often cease to menstruate, get morning sickness, start to lactate, suffer birth pains and gain weight, especially in their abdomen. In some cases, the abdominal swelling subsides during sleep. Human pseudopregnancy is not related to the normal ovarian cycle, as it is in dogs – it has even been reported in men.

Bitches' cycles are very different from women's, because they have discrete breeding seasons separated by six-month periods of sexual inactivity. They have just one reproductive cycle per breeding season. Each breeding season they develop one crop of ovarian follicles which all grow and ovulate at the same time, causing a single period of estrus, or heat. Then, if they have not become pregnant, they become sexually inactive again for another six months until their next season.

As far as reproduction is concerned, everything 'just happens' to a bitch. She ovulates spontaneously, forms her corpora lutea spontaneously and, most importantly of all, keeps these corpora lutea going whether or not any puppies are present. It is as if her body has assumed that she has become pregnant. Because she is not due to ovulate again for another six months, it does not make much difference to her ovaries if she is pregnant or not. Pregnancy lasts only nine weeks in bitches, and so any puppies will be born and weaned long before any future ovulations are due.

Dogs and their close relatives have taken the 'assume you're pregnant' system to its most extreme form. Developing puppies do not have to impose their will on their mother at all: there is no maternal recognition of pregnancy. Also, because the corpus luteum survives for the same length of time regardless of pregnancy, there are few hormonal differences between pregnant and barren bitches. This hormonal equivalence between pregnant and non-pregnant bitches explains why non-pregnant bitches exhibit pseudopregnancy. Their hormones change their behaviour in the same way as if they are pregnant: nesting, hiding puppy-like toys, lactating and even straining as if they are giving birth. Although other hormones are involved (prolactin, for example), the underlying cause of this pregnancy-by-default is the progesterone-secreting corpus luteum, which survives regardless of any puppies.

The 'assume you're pregnant' method may seem a cop-out, but if a mother doesn't need to know if she is pregnant, then why bother finding out? Still, surely life would be easier for the poor old mutt if she could avoid the exhausting effects of pseudopregnancy. There are two other reasons why dogs might use this strategy and both of them are linked to the social nature of these animals. Dogs are descended from wolves and these, like their domestic counter-parts, are extremely sociable creatures. Most adult female wolves live their lives in a stable pack group, and this probably goes some way to explaining their rather *laissez-faire* attitude to the start of pregnancy.

First of all, it is possible that almost all bitches can 'expect' to become pregnant whenever they ovulate because of the constant proximity of pack males. Why recognise pregnancy if it is almost a foregone conclusion? Furthermore, canines are extremely mater-nal creatures whether pregnant, non-pregnant or even male. Pet owners can hardly have failed to notice that while cats treat toys as if they are something they are trying to murder, dogs treat toys as if they are something they are trying to mother. A great deal of cross-parenting goes on within wolf packs, and bitches spend much of their time helping to rear pups of other, often closely related bitches. Because of this, it has been suggested that lactation in the barren bitch may have evolved as a way for her to contribute to the welfare of her infant cousins, nephews and nieces.

In other species, there are several variations on this doggy 'assume you're pregnant' theme, the first of which is called induced ovulation. A good example of this is the ferret. Ferrets have cleverly avoided the inconvenience of pseudopregnancy when they fail to become pregnant. Jill (female) ferrets have cheated the system by discarding one feature of bitches' reproduction: spontaneous ovulation. Rather than simply ovulating when their internal clock dictates, jills, not unreasonably, wait until they have been mated. The physical stimulation of mating causes the pituitary to release a burst of gonadotrophins to induce ovulation and the formation of the corpus luteum. Once the corpus luteum is formed then, as in the bitch, it will last for the same length of time regardless of whether a pregnancy has been established.

This system is economical because it stops jills from undergoing the hormonal changes of pregnancy when they have not been mated. However, as in the bitch, the female's decision as to whether she is pregnant is still not related to the presence of embryos. So while the ferret's new refinement allows considerable savings in time and energy, it still falls firmly into the 'assume you're pregnant' group. The intervention of the brain in ferrets' ovulation is a good example of how the brain allows important reproductive decisions to be influenced by environmental cues, in this case mating. Again and again we will see the brain interfering in reproduction when it seems to feel it can help, and this is probably the reason why the ovaries have become so dependent on the pituitary gland.

The ferret's system is not perfect, however, and female ferrets can suffer because of their mating-induced ovulation. If a jill is not mated for a long period of time, she accumulates large unovulated ovarian follicles that pump large amounts of estrogen into her system. This is a dangerous state of affairs because high doses of estrogen are toxic. Prolonged estrogen poisoning suppresses the bone marrow of unmated female ferrets and they often succumb to overwhelming infections, rather like human chemotherapy patients. So ferrets can die for lack of sex.

Another group of mammals, the marsupials, has developed the 'assume you're pregnant' system to a level of refinement exceeding that seen in other mammals. Marsupials' methods of reproduction are very different from ours, and in some ways they are superior. In particular, their ability to manipulate their pregnancy-initiation system has allowed them to use reproductive schemes impossible for other mammals. The characteristic feature of marsupials is that most of their infant development is supported by milk rather than a placenta. Whereas 'placental' mammals like us are reared to a fairly mature state by the placenta, marsupials use the opposite approach. They dispense with their placenta early, bear immature young and then do most of their rearing with milk. So, there is a deep division in mammalian reproduction, between those that have emphasised internal, placenta-maintained development and the others that have opted for a pouched, milk-maintained system.

Because their pregnancies are extremely short, many marsupials can use the dog-like 'assume you're pregnant' system. Hormonal changes can carry on the same regardless of the presence of embryos, because pregnancy is shorter than the interval between ovulations. The female has no need to anticipate pregnancy. Whereas dogs can use this system because their interval between ovulations is unusually long, marsupials can use it because their pregnancy is unusually short. Marsupial mothers can assume they are pregnant because any babies will be shot out well before the next ovulation is due. So the lifespan of the marsupial corpus luteum is often the same regardless of whether or not female marsupials are pregnant.

One of the most important implications of this marsupial system is that they are able to store new embryos in suspended animation until the time is right for them to be born. This allows them to rear several offspring of different ages at the same time. As soon as the older siblings have left their mother's pouch, or their mother's side, the tiny embryo within the uterus can be reactivated.

86

Although we have only looked at one of the possible strategies for controlling the start of pregnancy, we have already seen a variety of animals controlling their fertility to a remarkable degree simply by attuning ovulation, luteal formation and luteal destruction to selected environmental cues. The choice of reproductive system always comes down to what is important for the animal — whether she actually needs to know if she is pregnant; whether it would be easier to assume she is not pregnant if she has not mated; whether this new baby would be better stored until next summer when its older sibling will be weaned. Animals' decisions about their reproduction are important and nature has been resourceful in helping them decide wisely.

Strategy 2: waiting for the placenta

The rest of the mammalian strategies for starting pregnancy include females that, unlike dogs, respond differently depending on whether there is an embryo in their uterus. In these animals, the embryo must inform its mother of its presence. Yet the embryos in our second group achieve this maternal notification in a relaxed fashion. A mixed bag including cats and many rodents, this group of mammals seems to have trodden the 'middle path' of maternal recognition of pregnancy, and their embryos are in no undue rush to announce their presence.

The presence of these embryos becomes apparent to their mother some way into pregnancy when the placenta starts to make its own progesterone. Until this time, which can be a third or a half of the way into pregnancy, their mothers must assume the presence of their embryos and slavishly maintain luteal progesterone secretion. If the female members of this 'await the placenta' group subsequently turn out not to be pregnant, then they will have endured a pseudopregnancy. This pseudopregnancy, while not as long as a normal pregnancy, must last as long as it takes an embryo to form a placenta. Still, maternal recognition of pregnancy does happen in these species because the embryo does, albeit belatedly, announce its presence.

Just as straight 'assume you're pregnant' was not sufficient for ferrets, the most basic form of 'await the placenta' has also proved inadequate for many mammals, and some have evolved strategies to tune the system for their own reproductive demands. Cats, for example, have developed induced ovulation (like ferrets), so that they do not embark on a futile pseudopregnancy if they have not even mated. Feline copulation induces a surge of pituitary gonadotrophins that precipitates ovulation and formation of the corpus luteum. Mating does not automatically imply conception, but just like ferrets, cats save a lot of time by assuming that if they do not mate, they cannot conceive. Because they cannot ovulate until they have been mated, isolated queen cats spend much of their adult life with large follicles pumping out estrogen into their system. This is why queen cats spend so much of their time in heat, and tomcats spend so much of their life trying to make the most of this situation.

In comparison, rats really are most intelligent creatures, and their version of 'await the placenta' is one of their great evolutionary triumphs. Like cats, they have developed a system that allows them to avoid pseudopregnancy if they are not mated, but unlike cats they do not have to spend much of their life in heat. They achieve this feat by careful juggling of the usual three events – ovulation, luteal formation and luteal destruction. Female rats ovulate spontaneously, and if unmated they will do so every three or four days. Although ovulation is spontaneous, however, conversion of the follicle to a corpus luteum takes place only after mating. This corpus luteum will then survive for about twelve days until the placenta has to take over. At this point, any mated but non-pregnant rats will finish their pseudopregnancy and ovulate once more. By contrast, pregnant rats will go on to produce their pups around twenty-one days after mating. Rats have really hedged their bets: if they do not mate, they ovulate at frequent intervals; if they mate but do not conceive, they ovulate less frequently because they undergo a pseudopregnancy; if they mate and conceive, they will ovulate most infrequently of all.

The 'await the placenta' system can cope with all sorts of complications, and is clearly the best system for many mammals. Even in the rat, however, its efficiency is limited by the speed with which the embryo can grow its placenta. In many species this wait is simply too long. Mammals in a rush to conceive as quickly as possible had to develop a new system. These mammals have two choices – they must join one of our last two pregnancy-initiation groups, along with Harvey's beloved deer and ourselves.

Strategy 3: the magic message – interferons and the start of pregnancy

Red deer really are in a hurry, so they have to have a quicker pregnancy-initiation system than cats and dogs. But why the rush? As Harvey realised, the sudden synchronous entry of all the female deer in a herd into the breeding season means that all their calves will be born at the same time the next year. What is so special about that time of year?

Fortunately, red deer are just about the only animal in which both sides of the pregnancy story have been studied: their reproductive needs in the wild and the mechanisms of conception that address these needs. For the last three decades, the ecology of a population of feral deer on the small Scottish island of Rhum has been studied as intensively as that of any mammal population in the world. These feral deer have given us an unprecedented insight into how animals interact with each other and their environment to optimise their success, although they are unfortunately currently under threat from planned changes in land use on the island. They have taught us that wild deer are under tremendous pressure to give birth at a particular time of year, so that enough food is available to support the growth of their calves before their first hard winter. The greatest demands are placed on hinds not during pregnancy, but during the lactation that follows, so they aim to produce their calves just before the peak summer growth of grass. For example, it has been demonstrated that calves born late will not be able to grow enough before winter comes and are more likely to die. Conversely, early-born calves are usually lighter at birth and this is a disadvantage that they are not able to overcome before the winter either. So female red deer constantly walk a tightrope between life and death – simply conceiving at the wrong time can destroy their chances of producing any grandchildren.

As the social structure of red deer herds is designed so that a dutifully randy stag is present when required, it makes sense for the hind to get cracking, ovulate and form a corpus luteum spontaneously. After a hind mates at the start of the autumn breeding season, she must find out pretty quickly if she has conceived. Obviously, if she has conceived, then everything is fine. If she has not, she must know that conception has failed as soon as possible so that she can ovulate again. The longer she waits, the later her calf will be born, and we have seen how disastrous that could be. Everything depends on the corpus luteum she formed when she ovulated. If she has not conceived, then it must be destroyed in seventeen days. If an embryo is present, then it must save the corpus luteum, or perish. Deer, along with most other hoofed animals, are therefore 'corpus luteum protectors'.

The spontaneous destruction of that all-important corpus luteum in non-pregnant hinds is the result of a carefully choreographed hormonal ballet. By two weeks after ovulation, the uterus, which had been awaiting the arrival of an embryo that never came, changes slightly. It starts to develop receptors for the hormone oxytocin, which is made by the pituitary and the corpus luteum. In response to oxytocin, the uterus secretes another hormone, prostaglandin. This prostaglandin then makes the corpus luteum make even more oxytocin. This back-and-forward dance of mutual hormone stimulation between the uterus and the corpus luteum happens in several fitful episodes, and the resulting pulses of prostaglandin destroy the corpus luteum.

| Uterus | $\longrightarrow$ | Corpus luteum |
| Prostaglandin | $\longleftarrow$ | Oxytocin |

The whole system is on a hair trigger. A slight increase in the sensitivity of the uterus to oxytocin is all that is needed to set off a self-destruct programme for the corpus luteum. So it is clear what the deer embryo has to do: it has to stop this self-destruct programme from ever starting.

Red deer embryos release a single chemical to stop the destruction of the corpus luteum. First detected in sheep and cattle embryos, but now also in deer, this chemical is a protein called interferon. Deer embryos produce a lot of interferon while they are still drifting free in the uterus. Interferon stops the uterus from becoming sensitive to oxytocin and so jams the luteal self-destruct switch. *Voilà!* Corpus luteum persists, lots of progesterone is produced and pregnancy starts.

Many people have heard of interferon because it can be used to treat multiple sclerosis. A few years ago, interferons were being touted as the new panacea and, although they have been out of the headlines for a while, they are still being intensively studied for their medicinal properties. As it happens, the interferon of deer pregnancy is closely related to alpha- and beta-interferons made by cells when they are infected by viruses. In fact, before the discovery of embryo interferons, interferons were thought to be important only in fighting disease.

No one knows why interferon has acquired its new role in early pregnancy in hoofed animals. It is likely that deer's embryos once produced interferon for some other reason, perhaps to stop viruses colonising the pregnant uterus. Over millions of years, deer mothers may have started to detect this anti-viral protein and use it as a cue to tell them their embryo was there. There is no longer any evidence of an anti-viral role for embryonic interferons in deer, but that does not mean that they did not have such a role in the past. In fact, the special type of interferon made by deer and sheep embryos still retains many of the old anti-viral properties of its predecessors. Surprisingly, these cloven-hoofed mammal embryonic interferons have even been proposed for drug use in humans, as they may have fewer side-effects than human interferons. Suggested uses have included the management of lung cancer and multiple sclerosis.

The deer's method of arresting maternal sexual cycles by protecting the corpus luteum with an interferon has proved to be very flexible. Not only has it allowed different hoofed mammals to develop different reproductive strategies, but it has also allowed them to become finely attuned to succeed in their environment.

For example, the roe deer is able to hold embryos in suspended animation, just like many marsupials do.

Reproduction in deer even seems to be flexible enough to allow females to select the sex of their offspring on the basis of their social hierarchy. Evidence from wild deer has suggested that dominant hinds tend to produce male calves more often than submissive hinds, who produce more female calves. There is a fairly clear logic to why hinds should want to do this: social status is often inherited in red deer, and so dominant hinds tend to produce dominant calves. A dominant stag can produce perhaps thirty offspring a year, and so is likely to provide its mother with many grandchildren. The other side of this is that submissive stags often produce no calves at all. So, it makes sense for a submissive hind to produce a female calf, as most hinds produce one calf a year, even if they are submissive.

Recent work has suggested that the red deer embryo interferon may act on the brain as well as the uterus to inform its mother that it is pregnant. This may be how the brain (which 'knows' the social status of the mother) and the embryo (which 'knows' what sex it is) somehow reach a decision about whether to continue with pregnancy. With the advent of new techniques of reproductive medicine, many human couples are spending large amounts of money trying to have a child of the chosen sex. Deer appear to have been doing exactly that for millions of years, for free.

So, in Harvey's beloved deer, we have discovered an elegant network of different external influences 'made flesh' by the coordination of brain, womb and ovary. It has taken scientists centuries to unravel this tangle of hormones, but now we can see how inventive nature can be when the stakes are high.

Strategy 4: humans and the 'placental brain'

Humans are members of a group of animals that, like deer, need a quick decision – animals that cannot wait for pregnancy to run its course to find out if they have conceived. Time is of the essence in humans as much as it is in deer. In fact, time is more precious because of menstruation: deer embryos have about three weeks to inform their mother that they are present, but human embryos have less than two weeks to do the same thing. Although it is nearly a month from fertilisation until a woman's next ovulation, it is less than a fortnight until the date she has pencilled in for menstruation.

When it comes to the mechanisms underlying luteal destruction in non-pregnant women, I will recount no more elaborate hormonal machinations, no more Byzantine reproductive plots. Luteolysis in people is not boring; it is just not as well understood as luteolysis in deer. The main reason for this is that humans are inherently harder to study. Whereas luteal destruction in red deer involves communication between organs scattered throughout the body, most of the interactions that control the process in humans probably take place within the ovary. The chemicals involved (probably estrogen and prostaglandin) may not be released into the general circulation in large quantities, so we cannot measure them by radioimmunoassay of blood samples. Luckily, however, we do not need to know exactly how luteolysis works, as the human embryo has a relatively simple plan to prevent it.

The human embryo has a forceful approach to the maternal recognition of pregnancy. For a start, it does the exact opposite of what deer embryos do – instead of drifting around free in the uterus, it sticks to the uterine wall and digs itself in. This process is so rapid and aggressive that the embryo can form a considerable bulk of placental tissue as early as five to seven days after fertilisation. Compared to many mammals, this may seem like undue haste – especially as human babies have a rather leisurely pregnancy in most other respects – but this early invasion is a critical step in the embryo's bid for the security of a non-cycling mother. The early placental cells immediately start to pour out huge quantities of a hormone called human chorionic gonadotrophin, or hCG, and this is the human pregnancy recognition signal.

When it is first produced after ovulation, the corpus luteum is supported by gonadotrophin hormones made by the pituitary. In a non-pregnant woman, as time goes by and the future of the corpus luteum starts to look uncertain, this gonadotrophin support gradually starts to fade and eventually the corpus is destroyed.

93

The embryo's solution to this failure is to swing the balance back towards luteal maintenance, and it does this with hCG.

Because it is chemically very similar to pituitary gonadotrophins, hCG acts like a gonadotrophin. The effect of hCG is quite crude: unlike deer interferons, which selectively switch off a single critical part of the luteolytic mechanism, hCG simply overwhelms luteolysis. The human embryo is the control freak of the animal kingdom – it burrows into the uterus and starts supporting the corpus luteum itself. So, we can call our fourth group of mammals the 'corpus luteum supporters'.

So, hCG is more than just a pregnancy recognition signal: it also carries within it the hormonal drive to pregnancy. The early appearance of hCG means that designing pregnancy tests for women has been quite easy. There is so much hCG in the bloodstream that it spills over into the urine, and this is the most convenient place to look for it. The test kits that you can buy in shops are essentially the same as the ones used by your doctor: they all detect hCG. With sensitive pregnancy tests widely available, women can now know they are pregnant at around the same time that their baby is giving their corpus luteum the same news.

Yet hCG is not all good news. Of all the biological burdens pregnant women have to cope with, morning sickness seems the most unnecessary, even perverse. It is difficult to think of a reason for women to vomit incessantly during early pregnancy. A slight tendency to be more careful with what they eat would seem reasonable, but this could not explain repeated paroxysms of nausea. Even if we could argue a possible function for morning sickness, we would then have to explain how many pregnant women cope perfectly well without it, as well as why some women get morning sickness during some of their pregnancies but not others.

Morning sickness can be extremely debilitating. It is thought that Charlotte Brontë died of the condition, and if anyone was able to describe the horrors of morning sickness it was her – she described straining 'until what I vomit is mixed with blood' and much of her final days was spent in a 'walking delirium'. Although death from morning sickness is now rare, between 0.5 and 1 per cent of women with the condition need intravenous fluid therapy to keep going. Even so, perhaps thirty women are saved each year in Britain when their morning sickness becomes so life threatening that it is cured by a termination. Doctors and drug companies are reluctant to consider using anti-sickness drugs for the disease, as the last one used widely for the condition was Thalidomide.

Recently, a controversial theory was proposed suggesting a possible function for morning sickness. It is claimed that women frequently become most nauseous when presented with bitter foods, coffee being an excellent example (although my wife gave up tea for coffee while pregnant, just to be different). Bitter foods are usually plants containing 'secondary compounds,' which are poisonous substances designed to stop animals eating them. Popularised in a book, this theory asserts that aversion to bitter foods helps pregnant women to avoid poisonous plant chemicals that would harm their developing baby. Many scientists and doctors have reacted against this proposal, partly because there is no evidence to support it, and partly because they fear that it may stop women eating vegetables during their pregnancy. The author has even been pilloried, somewhat unfairly, for inflicting un-founded guilt on women who give birth to children with abnormalities after eating vegetables during pregnancy.

In defence of the theory, it has been claimed that miscarriage is less common in pregnancies affected by morning sickness, and that morning sickness is not associated with reduced birthweight of babies. Of course, all this may mean is that morning sickness is the sign of a healthy fetus, or that fetuses are not really affected by their mothers' morning sickness. Similarly, the fact that women who eat lots of vegetables during pregnancy tend to have larger babies has been used to attack the theory, but this could be because people who eat a lot of vegetables may have a better overall diet. As the debate stands now, there is little good evidence that either supports or disproves the idea that morning sickness is protective, and many people believe instead that morning sickness is an unfortunate side-effect of hCG.

An increasing body of evidence is now beginning to implicate hCG in morning sickness. The first piece of evidence comes from women with placental tumours – they often have increased amounts of hCG in the circulation, presumably because of the excessive bulk of placental tissue in their uterus. These women often also show signs of overactivity of the thyroid gland in the neck – not usually enough to cause obvious symptoms, but enough to show up on blood tests. Also, when researchers looked at hCG and thyroid function in women with normal embryos, but afflicted by the most severe form of morning sickness (given the rather charming name of hyperemesis gravidarum), they found a link between blood hCG levels, blood thyroid hormone levels and the severity of vomiting. This association does not mean that hCG actually causes thyroid stimulation or morning sickness, but there are good reasons why this might be the case.

Pituitary gonadotrophins and hCG have similar effects because they are chemically very similar. There is another hormone that is very similar, and this is thyroid-stimulating hormone. This hormone is released by the pituitary and, as its name suggests, it controls hormone release from the thyroid gland in the neck. Thyroid hormones control the body's metabolism, so it is possible that hCG causes morning sickness by interfering with the thyroid-stimulating hormone/thyroid system. Indeed hCG can bind to the thyroid, although it is not as effective as pituitary gonadotrophins – if the placenta released pituitary-like gonadotrophins rather than hCG, perhaps morning sickness would be ten times worse. This may even be why hCG is slightly different from other gonado-trophins – to reduce its unwanted effects on the thyroid.

There is one final, inexplicable twist in the morning sickness saga. Not only do different babies induce different levels of severity of morning sickness in their mother, but this difference even occurs between siblings (my sister and myself, for example). There is now good evidence that children who cause worse morning sickness in their mother are more likely to like the taste of salt once they are born. Although reasons such as vomiting-induced dehydration have been proposed as a cause of this altered salt preference, nobody really knows why it exists. Could it even be a useful 'reason' for morning sickness?

The controversial story of hCG brings us to the end of our quest for the mechanisms of conception. We have seen how, for a successful pregnancy to be established, the developing embryo must address the fundamental problem of its mother's dual-function reproductive organs by switching her from a cyclical, sexual being to a stable, pregnant one. We also saw how the reproductive cycle is a graceful series of sequential strides towards ovulation and the formation of the corpus luteum, and that it is this latter organ which holds the key to pregnancy.

Some female mammals (bitches) simply wait to see if any babies are born before they decide to destroy the corpus luteum and ovulate again. Others (queen cats) are in slightly more of a hurry and can only wait as long as it takes for a placenta to form and start secreting its own progesterone. A third group of mammals (red deer hinds) allows only the briefest of opportunities for its free-floating embryos to release interferon to prevent luteal destruction. The fourth group includes us: human embryos invade into the uterine wall and start to direct pregnancy themselves.

Also, it seems that at every available opportunity, the brain has been invited to collaborate in starting pregnancy. As the reproductive system's link to the outside world, the pituitary gland feeds in information about the time of year, the mother's lactation status, whether she has mated or not and even what sex of baby she wants. Humans have taken this trend to its limit, with the placenta releasing a modified form of a pituitary hormone as the pregnancy recognition signal itself. This 'placental brain' is surely as intertwined as mind and body can get. The similarity to Harvey's duality of 'conception in the mind' and 'conception in the womb' is remarkable.

And, of course, our human embryo was in the greatest hurry of all. It has secured its future within a fortnight of its creation and its mother has acquiesced and accepted its presence. It now has some time and a safe place in which to spend it, but of course it cannot rest. A gargantuan task lies ahead, one which will take up most of the next seven weeks. Having made itself a home, the embryo must now make itself.

○ ○ ○ ○ ○ ○ ○ ○

How is a baby
put together?

○ ○ ○ ○ ○ ○ ○ ○

CHAPTER 3

Making babies

○ ○ ○ ○ ○ ○ ○ ○ ○

You remember strange things from your childhood. I suppose I was five or six years old, which means that my sister was eight or nine. Sarah was a fairly strong-willed and persuasive character (she now works in advertising) whereas I was evidently a fairly malleable individual (I am now a scientist). It was not really a fair fight. One thing is certain, however: I am in no doubt that Sarah, in her latest attempt to test my gullibility, did not realise that she had accidentally reformulated the erroneous thesis of a major nineteenth-century German evolutionary biologist.

Sarah had a wonderful book, of which I was extremely jealous, and which by various subterfuges has now fallen into my possession. The *Living World of Animals* was as impressive in its thoroughness as it was in its sheer weight. I managed to monopolise it quite successfully, and I remember spending much of my time with my nose buried in the fauna of temperate deciduous forests and the fishy wonders of the abyss. I must admit that the last chapter, about evolution, did not interest me much – perhaps I had never really thought there was an alternative to the idea that animals changed over time and I simply accepted it.

In a rather old-fashioned way, the book depicted the 'progression' of evolution from single-celled amoebae, through backbone-less animals, fish and reptiles to mammals and thence on to ourselves. In fact, such 'great chain of being' depictions of the evolutionary process are still quite common in children's books. Perhaps it is the reassuring impression of perfection that they confer on our own species that has made them so enduring.

Anyway, for some reason my sister hit on the idea of seeing if she could persuade her younger brother that he had actually gone through all the scaly and furry stages of evolution before he was born.

'Surely you must at least remember when you looked like *that*,' she would say, pointing emphatically at a picture of a monkey trailing along behind some caveman-like chap. 'That would have been the last bit before you were born. Don't you remember being furry like that?' I had to say that I did not really remember being furry at all, but who was I to say that she was wrong? After all, the animals must have been arranged on the page like that for some reason. I cast my mind back as far as it would go, but it would not cast back much further than when I was two or three. I think my first memory was a bonfire – certainly nothing as exciting as being furry and having a tail. A tail would have been cool. I was sure I would have remembered that.

But Sarah was adamant, so what could I say? Perhaps I was just an inadequate who had forgotten about his time as a monkey. I thought it might be best to claim that I knew what she was talking about and then go away and try and work the whole thing out in private (as I said, I am now a scientist). Nothing was worse than to admit ignorance, so I played along. 'Oh yes!' I said. 'Perhaps I do remember a little bit.' This seemed to satisfy her – after all, what sort of moron can be talked into remembering something that never happened? She was happy.

During my teens I was taught about animal evolution and embryonic development as separate entities, but the two remained strangely linked in my mind. Could my sister have unknowingly been right? Is it just a childish fiction that we could each be cells, worms, fish, reptiles, small furry mammals and then monkeys before we are finally born as little people? Is our development in the womb some sort of replay of our evolutionary history? Could this have something to do with baby monkeys looking a lot like baby people? And what about tadpoles – are they not frogs' developmental homage to their fishy ancestors? The answers to these questions are still not fully known, but they have a lot to tell us about how a human baby forms. The idea that we replay our evolutionary history as we develop in our mother's womb was first proposed in the mid-nineteenth century by the German embryologist Ernst Haeckel, as his proudly titled 'biogenetic law'. Although wrong, this 'law' has been the backdrop of our studies of evolution and embryonic development ever since.

Although we have got to Chapter 3, the embryo whose story we have been dutifully following is still just a simple ball of cells. How do you make a baby out of that? Surely not by making a fish or a monkey first?

Haeckel's embryos

Ernst Haeckel, Professor of Comparative Anatomy at Jena, was a true original. By a strange Teutonic mixture of idealist mysticism, amazing scientific productivity and the social graces of a steam-roller, he came to dominate the field of embryology in the late nineteenth century. Obsessively anti-religious, he espoused a union of science, philosophy and religion that seems rather overambitious today. Inspired by Goethe, Haeckel saw the world as a series of phenomena that could be progressively unified into an all-encompassing whole. To him, the dualities of body and spirit, matter and energy, God and nature were mere façades covering single entities beneath.

This anti-dualism came to be called, rather inelegantly, 'monism', and Haeckel, it has to be said, was prone to drone on about it at great length. Hardly a paragon of literary clarity at the best of times, his diversions into this philosophy are frankly almost unreadable. With hindsight, however, the monist concept has proved to be rather successful. For example, twentieth-century physics has shown that matter and energy are in fact interconvertible and the study of the role of the brain in disease has started to chip away at the old idea of separate mind and body. As for God and nature, I will leave that one up to you.

The apparent success of monism was probably just an inevitable result of the way people approach problems. If they do not understand something, then they often try and break it up into smaller, simpler bits that they do understand. At some stage they can then reassemble these pieces to see the whole phenomenon as it existed before they broke it up. The nineteenth century was the pinnacle of this sort of intellectual dissection: the obsessive cataloguing of the Middle Ages had been replaced by incisive and enlightened 'reductionism'. The mechanisms of the cosmos had been dismantled for detailed inspection, and so it is perhaps not surprising that a philosophy of reassembly should seem, at first sight, to be so ahead of its time. If you come in halfway through this process, then of course you will find people thinking about lots of different intellectual strands that can eventually be reunited into their original grand design. Unfortunately for Haeckel and the monists, there is a problem with their approach – it is based on nothing more than philosophical gut feeling. What if two entities really are separate, or what if they overlap a little, but not much?

It was this mistake that led Ernst Haeckel to propose the greatest mistaken theory in the history of biology. In fact, I would say it is the greatest erroneous idea in all science, not because it is more spectacularly wrong than any other, but because it is wrong in such a deliciously interesting way. According to this theory, every animal re-enacts its evolutionary history as it develops, resembling ever more recent ancestors as it approaches maturity. Thus humans were claimed to look like single-celled animals, worms, fish and monkeys at progressively more advanced stages of pregnancy. This claimed correspondence of embryonic development to evolutionary history was not simply a coincidence – it reflected how animals evolve into new forms by adding new stages to their embryonic development. Thus, by Haeckel's 'law', evolution and development became two sides of the same process: a powerful monist argument.

Although long since discarded, Haeckel's self-proclaimed biogenetic law has proved to be an amazing catalyst to the study of evolution and embryonic development. It could never have had such a profound effect had it been correct. Indeed, if it had been correct, and evolution and embryonic development had proved to be directly equivalent processes, then the study of both would have been consigned to the backwaters of the natural sciences. Instead, many of the greatest minds of the twentieth century have applied themselves to trying to understand the related, but distinct, processes of evolution and development. In fact, the most interesting feature of Haeckel's erroneous law is that it seems to work disquietingly often. It is always interesting when a law that you know is wrong has such apparent predictive power. What could be more challenging than a theory that is almost right?

Even before Haeckel, the nineteenth century had been a time of intellectual ferment in biology as naturalists argued over the nature of change in animal species. In his own writings, Haeckel was eager to emphasise the importance of the contribution of his first great influence, Jean-Baptiste Lamarck, Professor of Invertebrate Zoology at the Paris Natural History Museum. Lamarck has been dealt a raw deal by posterity: instrumental in establishing the idea that animal species could change over time, or 'evolve', his name has become synonymous with scientific error simply because the mechanism he proposed for this change is now thought to be wrong. In his *Philosophie Zoologique* of 1809, Lamarck suggested that animals could, by their own efforts, alter the nature of their offspring. A good example is the giraffe. Did ancestral giraffes instil a longer neck into their offspring because they actively stretched for ever-higher branches? Could environment and experience contribute to the sculpting of the next generation? Now that Darwin and Mendel have shown us how animals cannot change the genes that their offspring inherit, Lamarck's idea may strike us as silly. Yet at the time he proposed it, there was no evidence to disprove it, or to suggest that it was inherently flawed. It seems rather harsh that this semi-error should have consigned Lamarck to such scientific ignominy.

Reading Darwin's *On the Origin of Species* in 1862 was clearly a seminal moment in Haeckel's career. He was immediately impressed by the simplicity of Darwin's theory of how evolutionary change actually takes place. Like Lamarck, Darwin accepted that animal species change over time, but, unlike Lamarck, he proposed that the tiny changes that animals accumulate are not imposed on offspring by their parents, but that animal species are passive entities slowly changed into novel forms by 'natural selection'.

Darwin's argument went something like this. There is variation between the individuals within any population of animals. Choose any characteristic, such as neck length in giraffes, and you will find a spectrum of that characteristic within the population (from individuals with shorter necks to ones with longer necks). If external influences so dictate (such as taller trees), then animals at one end of the spectrum (the long neck end) will be at an advantage over their contemporaries. They will get to be fit, fat, fertile individuals who will produce more offspring, and these offspring will inherit the advantageous characteristic. After many generations, that characteristic will start to predominate in the population and the species will have changed: their necks will have got, on the whole, longer. No changes have occurred within any individual animal: all the animals did was provide variation on which natural selection could work.

Haeckel leaped on Darwin's ideas with alacrity, and soon became one of his greatest supporters. As a result, a distinct German school of evolutionary biology, 'Darwinismus', was established with Haeckel at its centre. Unlike Darwin, who was judiciously cautious about his ideas, Haeckel was determined to establish and extend Darwinism so that it was beyond contradiction. This enthusiastic support led to the first meeting between the two men in 1866. With characteristic linguistic flourish, Haeckel recast Darwin's austere aspect as 'the broad shoulders of an Atlas that bore a world of thought: a Jove-like forehead, as we see in Goethe, with a lofty and broad vault'. Darwin, although rather overcome by Haeckel's bombastic and exhausting nature, was still forth-coming with his own, rather more measured felicitations: 'I have seldom seen a more pleasant, cordial and frank man.'

Haeckel's breakneck-speed extension of Darwin's theories came in three parts. The first was to speculate about the origin of life, a topic that Darwin wisely seemed determined to avoid. The second was to approach the question of human origins. Darwin realised that the implications of his ideas for human origins would be rather unpalatable to the Victorian mind and pondered long and hard before publishing *The Ascent of Man*. Driven by his iconoclastic monism, Haeckel had no such reservations. Back in England, however, Darwin was not so much disturbed at Haeckel's treatment of his ideas as completely baffled by the sheer enormity and incomprehensibility of Haeckel's writings: 'I am sure I should like the book much, if I could read it straight off instead of groaning and swearing at each sentence.'

Haeckel's third extension of Darwinism is the idea for which he will always be remembered, and the one of interest here. Haeckel was eager to marry the two main threads of his intellectual life into one monistic whole. Savants had discussed the relation between embryonic development and animal evolution for some time, but the debate reached a crucial point in Haeckel. He had spent much of his career studying the embryological development of a bewildering variety of different animals, and now he had an evolutionary context in which to reassess his findings.

With characteristic boldness, Haeckel made the intellectual leap that his monistic philosophy demanded: he claimed that species evolution and embryonic development are essentially the same process. When he looked at early mammalian embryos, he believed he was watching the early stages of these animals' evolution, and because of this he surmised that animals actually re-enact their evolutionary history during their embryonic development. He claimed that before a mammalian embryo starts to look like a mammal, it looks like a reptile and before that it looks like a fish. Similarly, a reptilian embryo goes through a fishy phase before it takes on reptilian form. A major piece of evidence to support his remarkable claim was the appearance of gills at an early developmental stage in fish and all their descendants:

By a tenacious heredity these gill-clefts, which have no meaning except for our fish-like aquatic ancestors, are still preserved in the embryo of man and all the other vertebrates. Even after the five vesicles of the embryonic brain appear in the head . . . the foetus is still so like that of other vertebrates that it is indistinguishable.

To Haeckel, this 'tenacious heredity' suggested a guiding mechanism. To him, the playing-out of evolutionary history during embryonic development was real – it was a direct reflection of how embryonic development is a direct recapitulation of evolution. By claiming this, Haeckel was implying that development is equivalent to evolution. Development and evolution were no longer distinct. Instead, they had been fused into a single entity and this, of course, appealed to Haeckel.

I established the . . . view, that this history of the embryo (ontogeny) must be completed by a second, equally valuable, and closely connected branch of thought − the history of the race (phylogeny). Both these branches of evolutionary science are, in my opinion, in the closest causal connection; this arises from the reciprocal action of the laws of heredity and adaptation; it has a precise and comprehensive expression in my fundamental law of biogeny.

The biogenetic law was potentially extremely powerful because it worked both ways. If true, it meant that rather than digging through myriad fragmented fossils, biologists could infer the sequence of evolution of any animal simply by observing its embryonic development (the 'Law of Correspondence'). Conversely, animals' programmes of embryonic development could be interpreted as the end result of millions of years of accumulation of new developmental instructions tacked on to the end of the developmental programmes of their ancestors (the 'Law of Terminal Addition'). Hence, embryos acquire new features in the order that their ancestors evolved them: development recapitulates evolution. Fish became amphibians by compressing their fishy developmental programme into the first part of embryonic life (the 'Law of Truncation') and adding a new froggy addendum afterwards. Reptiles then added on another tier of instructions and finally mammals wrote the final chapter. According to the biogenetic law, the fishy bit of development is still there in humans − it is just squashed up at the beginning of embryonic life.

The idea that embryonic development might recapitulate evolution was attractive, and clearly it obsessed Haeckel. Yet do babies really replay their evolutionary history as they develop in their mother's womb? Haeckel was well placed to look for evidence to back up his theory and this he proceeded to do.

Perhaps the most famous example of his evidence is a drawing that has been reprinted again and again in school biology textbooks. It shows embryos of several different vertebrate species at similar stages of embryonic development. Early on, the embryos of fish, amphibians, reptiles and birds look virtually identical and only later do they gradually acquire their distinguishing features. And, true to form, they all look just like tiny fish at the beginning. Haeckel even proposed that the development of animals from single-celled fertilised eggs had direct parallels in the origin of all many-cellular animals from single-celled organisms. What better proof could there be?

Even as it was formulated, however, there were cracks appearing in Haeckel's biogenetic edifice. Some of the problems were largely philosophical. Despite his love affair with Darwinism, it is still possible to see relics of Lamarck's evolutionary mechanism in Haeckel's writing. Haeckel's evolution seems imbued with a Lamarckian sense of purpose and progress that Darwin might not have recognised. Did this purpose imply an aim and did this aim imply a degree of responsibility? It was this sense of active involvement of animals in their evolutionary progress that eventually allowed value judgements to be applied to the evolution of different human races. Dangerously, this idea could be extended to support the existing chauvinism that each race could be assigned to a different level of progress or perfection, presumably reflecting each race's 'distinctively' industrious or lazy nature. The belief that Europeans are the advanced descendants of non-Europeans was common in Haeckel's time, and his concept of evolutionary purpose seemed to confirm this conceit. Indeed, his ideas were later to be the seeds of some of the more unsavoury parts of Nazi philosophy. In his book *The Riddle of the Universe*, he wrote that non-Europeans are 'physiologically nearer to the mammals – apes and dogs – than to the civilised European. We must, therefore, assign a totally different value to their lives.' Even worse, as part of his omni-unifying philosophy, which maintained that even politics and biology are two sides of the same coin, he even went as far as to arrange the world's religions into an evolutionary hierarchy as well. When he places Judaism as an intermediate between 'savage' paganism and 'advanced' Christianity, you can almost feel the logic of the Final Solution forming around his words.

There were also scientific problems with Haeckel's biogenetic law. In fact, it is surprising that the 'law' held sway for so long, especially as a more general and probably more correct theory had already been expounded by Karl Ernst von Baer, Professor at the National Academy of Sciences at St Petersburg. Earlier in the nineteenth century, von Baer had speculated about the link between evolution and development, and pointed out that anatomical features common to different groups of animals often develop early in their embryos, whereas characteristics that distinguish them from each other often form later. This idea of each species progressively diverging from its related species throughout embryonic development is probably rather nearer the mark than the biogenetic law. In von Baer's system, early human embryos are not fish, they are like embryonic fish. This is a more 'egalitarian' system – it does not, for a start, carry such a sense of evolutionary progress. The gill-like structures of fish and human embryos are not gills, they are gill-precursors in fish and bits-of-the-head-precursors in humans. They happen to have retained a roughly equivalent place in embryonic development because they probably evolved from the same original structures. Von Baer's idea is much looser than Haeckelian recapitulation and is an observation, rather than a law.

Haeckel revisited

The recapitulation controversy rumbled on into the twentieth century, but new insights into the link between evolution and development were to come from an unexpected source. Not being an evolutionary biologist, Walter Garstang approached the problem from a different direction. He was Professor of Zoology at the University of Leeds and his real interest was the larvae of marine organisms. Having worked on a diverse range of larval forms that metamorphose into an equally diverse range of adult animals, Garstang found example after example of embryonic development that clearly violated Haeckel's biogenetic law.

Because his embryos lived independent lives in the ocean, rather than being nurtured within their mother, it was obvious that natural selection acted on them as much as it acted on adults. The idea that natural selection could act on embryos was a clear challenge to Haeckel's theory. Haeckel's claim that evolution involves modification of the end of animals' developmental processes is directly contradicted by any animal for which natural selection acts on the embryo rather than the adult.

Garstang extended Darwinism because he realised that there *is* a link between evolution and development, but this link is far more subtle than Haeckel's direct equivalence. Garstang proposed that the story of the evolution of animals is the story of the evolution of their developmental programmes. Evolution and development must be strongly linked, but there is no reason why recapitulation should be that link. Novel features could be introduced at the beginning, in the middle or at the end of embryonic development, as long as the embryo could still function. He also saw no reason why evolution could not progress by loss of old features as much as by acquisition of new ones. Indeed, he proposed an alternative to recapitulation, 'paedomorphosis', in which infant, juvenile or even embryonic characteristics of ancestors are retained into adulthood in their descendants. Remarkably, many of the features of adult modern humans have been suggested to be infantile features of our primate ancestors that we have retained into adult life, such as our flat faces and high domed foreheads.

The ghost of Haeckel still haunts biology, and great minds have taken it upon themselves to exorcise it. Modern evolutionary biologists have wrestled further with the evolution–development link until we now have a multi-part model of how developmental programmes can evolve. It now seems that natural selection can change any stage of embryonic development and, although it often alters the end of development rather than the start, this is not a hard-and-fast rule as Haeckel believed.

Because of this flexibility, any part of an animal's developmental programme can be put to new uses. New structures are often made by adaptation of old ones rather than being developed from nothing. So related animals (for example, humans and salmon) frequently use the same basic structures (gill bars) to make different final bits of anatomy (human head skeleton and salmon gills). There is often no need to make the original structures in a different way in the two related groups, and as a result related species often share large sections of their developmental programmes. They do not have to; they simply do because of historical accident. This 'evolution by historical accident' is the reason why recapitulation often seems to work: related animals often retain common developmental processes, even if they put them to different uses later on in their embryonic development. Embryonic development is like a salvage yard: why make a new gill bar when you can just alter an old one?

We now have not one, but several possible 'laws' that can explain how developmental programmes evolve to produce different animals. The first is, of course, Haeckel's terminal addition of features at the end of the programme. Second, we have seen that, if done carefully, new instructions can be inserted earlier on in the programme. Third, we have Garstang's paedomorphosis, in which juvenile characteristics are retained into the adult. Fourth, adult features can be progressively exaggerated. Fifth, there are some special ways in which adult body size can change without altering body shape. As you can see, the list is now rather long and what seems to happen in the real world is that any combination of these processes can occur simultaneously. Your head can paedomorphose while your legs recapitulate.

Recent studies of development in flies and mammals have led to the most striking examples of salvage and reuse of ancient developmental mechanisms. One of the greatest gifts of the modern revolution in molecular biology has been the insight we have gained into the mechanisms underlying embryonic development – the 'programme' itself. Working on the fruit fly, researchers have unveiled an intricate network of inter-related genes that coordinate the formation of the body plan.

The homeotic genes are especially important. These genes play a central role in subdividing the fly's body into a series of segments, and then making these similar segments become different by inducing them to grow wings, or legs or mouthparts (homoios means 'alike'). In the late 1980s, scientists started to look for evidence of homeotic-like genes acting during development of mammals, and to their surprise they found a set of extremely similar genes switched on in patches of the mouse embryo, apparently to establish aspects of their body plan in the same way they do in flies. The common usage of this family of genes by animals as distantly related as flies and mammals is amazing, especially when you consider that the common ancestor that bequeathed these genes to them was probably some tiny worm-like creature. Despite their distant relationship, however, both groups have used their inherited homeotic genes for exactly the same purpose: arranging which bit of body goes next to which. On the embryonic building site, nothing, it seems, is too old to salvage.

The discovery that so many different animals use such similar genes to control their embryonic development is a sign that all animals use essentially the same mechanisms to coordinate the construction of their bodies. Although making a baby human and making a baby fly seem like very different and very complex processes, they both boil down to the same two things – telling cells where to go and telling cells what to be. After all, a baby's finger becomes a finger because the cells that made it were told to go to the end of the arm and make bits of bone, skin, muscle and nail. This essential simplicity of embryo construction is probably why all animals use the same tricks to form themselves.

There are only really two ways that an embryonic cell can be told what to do. First of all, cells can 'know' what they are meant to be because of their ancestry – if they are descended from a particular lineage of cells. In humans, this 'inheritance' method is probably more important towards the end of embryonic develop-ment. For example, at some stage, a group of cells is set aside to make a lung, and once these cells are assigned their fate, their descendants must all become lung cells – they have no choice. Early on in human development, cells have a much more uncertain future and they have the potential to become lots of different parts of the body. In many animals, the allocation of a particular fate to a set of embryonic cells is signalled by switching on a particular homeotic gene.

Another way that embryonic cells can be told what to become is by the chemical signals that drift through the early embryo. A good example of this is the *bicoid* protein that tells the fly embryo which end is its head end and which is its tail. There is more *bicoid* at the front of the maggot and less at the back, and each section of the developing fly works out whether to produce eyes, legs or guts on the basis of how much *bicoid* it has. *Bicoid* is inserted into one end of the egg by the embryo's mother, but the embryo soon starts to make its own chemical signals to orientate different parts of its developing body. This process also coordinates the fabrication of much of a human baby: how one side of the hand knows to make a thumb and how one part of the head is told to make an eye. We are all formed by the interplay of thousands of carefully placed chemical messages that waft through us when we are less than a few millimetres long.

The story of Haeckel has one final twist. It now seems that all is not well with his original evidence for his biogenetic law. Recently, a team of embryologists tried to recreate his grand comparison of vertebrate embryos, and found that the embryos simply do not look as similar as Haeckel would have had us believe. It seems that he 'tweaked' his data to fit his theory. Today this would be called fraud, but let us be generous and call it 'muddying the waters' between the observation, interpretation and representation of nature. If we do this, then the whole recapitulation episode is ennobled into an instructive lesson in the history of science. Perhaps that would be the most appropriate way to leave the most inspirational error in biology.

Making a baby

By about fourteen days after fertilisation, the human embryo has tunnelled its way into the wall of the uterus and ensured its future by arresting its mother's sexual cycles. Meanwhile, it has embarked upon a demanding programme of development, forming the basic structures that will produce the baby and the placenta and membranes that protect and feed it.

About three days after fertilisation, the embryo is still a simple ball of cells stuck inside the zona pellucida. Amazingly, in seven short weeks the embryo will be recognisable as, well, a very small baby. This change in external appearance is a sign that most of its internal organs have been almost completely formed within the same fifty days. Of the forty weeks of pregnancy, almost all the crucial stages of body formation are completed in the first seven – after that, the baby just grows. By the end of those first seven weeks, the baby will also have decided what sex it is going to be and will even have made considerable provision for producing its own children.

Around the fourth day, the embryo makes its first real developmental decision. Two different cell populations start to be committed to different destinies in the future embryo. A pocket of fluid forms within the ball of multiplying cells and this pocket enlarges until the whole embryo becomes a fluid-filled cell-lined sac, called a blastocyst, at around the fifth day. The outside of the sac is called the trophoblast, and stuck on the inside of the sac is a bundle of cells called, rather unimaginatively, the inner cell mass (Figure 2(a)). This is the first sign of segregation of discrete cell types within the embryo: the trophoblast and inner cell mass have very different futures ahead of them. The trophoblast will form the outer layer of the baby's placenta and membranes, and the inner cell mass will form the rest of the placenta and all of the baby.

The commitment of different cells to their respective fates is an indication of things to come. This process will happen again and again over the coming weeks until the body has been subdivided into its myriad component parts. The fact that this primal differentiation step is between the trophoblast and the rest of the embryo tells us a great deal about how human and mammal pregnancies are organised. The trophoblast is to be the embryo's contact with the source of its nutrients, its mother. For most non-mammalian vertebrate embryos, the main source of nutrition is the yolk, and so these embryos establish links with their huge yolk sac instead (humans do have a yolk sac, but it is relatively unimportant). So the trophoblast is especially important in mammals and it forms almost at the expense of the embryo, which develops slowly in mammals compared to most other vertebrates.

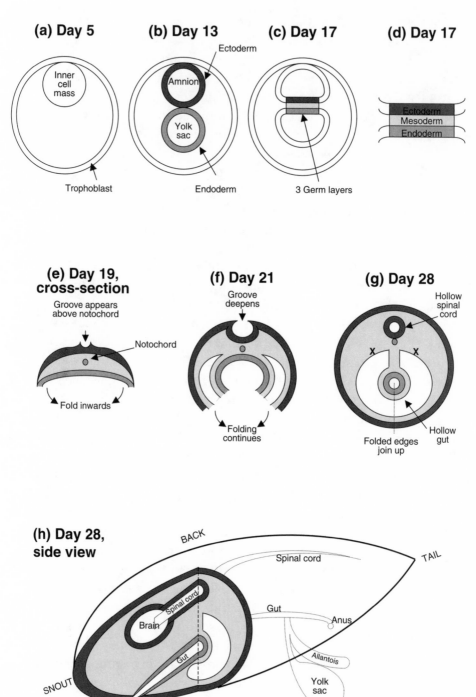

(i) Day 32, side view

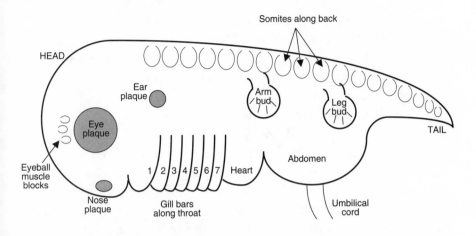

Figure 2. The formation of the human body plan in the first five weeks after fertilisation. Figures (a) – (d) illustrate how the embryo separates into trophoblast and inner cell mass, and that the latter subdivides further into the endoderm, mesoderm and ectoderm that will form the baby. In (e) – (g) these layers fold up to form the main structures of the human body, shown in cross-section (X denotes the position of the future kidney). By day 28 (h), the embryo has a distinct gut and nervous system, and by day 32 (i), the somites, gill bars, limbs, eyes, ears and nose can all be seen from the outside of the embryo.

Already we have hit on a problem that would, or should, have panicked Haeckel. We have reached the first act of cell differentiation, and already mammals are doing something different from fish. Forget gills – this is really early. According to Haeckel's biogenetic law, the processes going on at this early stage should be indistinguishable, even between species as distantly related as humans and fish. However, the rapid growth of the trophoblast in mammals is a good example of a new part of the programme that has been added at the beginning rather than at the end of development.

Apart from its job as the interface between mother and baby, the trophoblast is also medically important because it can also cause an unusual condition called hydatidiform mole. A 'molar' pregnancy usually starts normally, but uterine bleeding often starts a few weeks into pregnancy. Molar pregnancies are often larger than normal ones, but ominously, no fetal heartbeat can be detected. Despite this, the pregnancy continues to produce hCG as if nothing was wrong – in fact, these pregnancies frequently produce more hCG than normal ones. Hydatidiform moles are usually miscarried, but sometimes they must be removed artificially. Instead of an embryo, the uterus is filled with a mass of trophoblast tissue resembling a bunch of grapes.

These all-trophoblast-no-embryo pregnancies account for between one-twentieth and one-fifth of 1 percent of all pregnancies in the developed world, but they have been reported to be as common as 1 per cent in some developing countries. The cause of moles has been explained by their unusual genetic make-up – apart from their mitochondria, their genes are all inherited from the father. Each cell in the mole has two sets of genetic material, just like normal cells, but moles probably arise when a fertilised egg somehow loses its maternal genes, and ends up with two sets of paternal genes. Most often, this is due to the genes from the sperm being duplicated, but moles may also result from the simultaneous entry of two sperm into an egg.

In some ways, hydatidiform moles are the opposite of the parthenogenetic embryos we looked at in Chapter 1. Whereas parthenogenetic embryos contain two sets of genes from the mother, moles contain two from the father. And just as all-maternal embryos are not able to produce a normal placenta, all-paternal embryos seem to be able to make nothing except placenta. Nature obviously never intended us to have just one parent.

Making even more babies

All this time, the embryo has been trapped inside the same zona pellucida that enclosed the egg. Now that the embryo is a blastocyst, it must hatch out of the zona pellucida so that it can grow and get the nutrients it needs. In humans, hatching from the zona is crucial because the blastocyst must be free to implant in the wall of the uterus so that it can signal its presence to its mother, as we saw in Chapter 2. Also, the embryo has to escape the zona so that it can grow – in nine months' time it will be over a billion times larger.

The emergence of the embryo from the zona pellucida is a crucial time because it is at this stage that so-called identical twins are probably formed. These are different from non-identical or fraternal twins: fraternal twins are produced when a woman ovulates twice and both eggs are fertilised; a better name for these twins would be 'double-egg' twins. Double-egg twinning is thought to be a partially inherited characteristic – it can run in families. On the other hand, 'identical' twins are derived from a single fertilised egg, and as far as we know, these 'single-egg' twins do not run in families.

Although single-egg twinning appears to be fairly random in humans, it can be a tightly controlled, inherited process in some mammalian species. For example, some species of armadillos always produce a set of single-egg siblings at each pregnancy. Nine-banded armadillos always have single-egg quadruplets.

Single-egg twins can form in a several ways. First, the embryo can split into two, maybe as it crawls out of the zona. The two half-embryos then go their separate ways and implant at two different sites in the uterus, rather like double-egg twins would. Because of this, each has its own placenta and membranous bag containing it (Figure 3(a)). A second possibility is that the outer trophoblast layer of the blastocyst may stay intact, but the inner cell mass can split in two, resulting in two embryos contained in their own little bags of fluid, but enclosed by the same outer trophoblast membrane. Because they are retained within the same membrane, these single-egg twins must implant together (Figure 3(b)). The third option is a lot like the second, except that the splitting into twins occurs later and the twins share all their membranes, rather than just the outer trophoblast. Because of this, they actually float around together in the same sac of fluid (Figure 3(c)).

A common misconception about single-egg twins is that they are identical. If we are fully to appreciate the identity of 'identical' twins, it is essential to realise that they are never perfect replicas of each other. Parents of single-egg twins often try to ensure that the

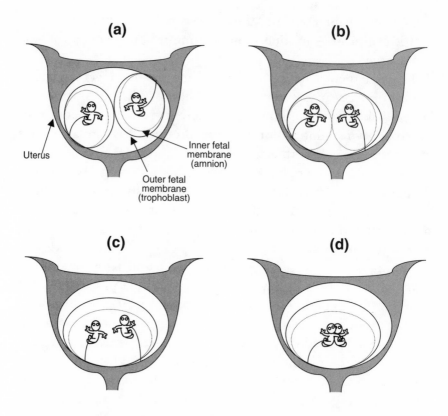

Figure 3. Different forms of single-egg or 'identical' twinning. In (a), the twins lie in completely separate sets of membranes, just like double-egg or 'fraternal' twins do. Alternatively, single-egg twins can share the outer (b) or both the inner and outer (c) membranes. Rarely, the twins' bodies may fail to separate fully, leading to conjoined twins (d).

twins grow up as two independent characters – by dressing them in different clothes, deliberately emphasising any differences they notice and this is probably a good idea. Fortunately, these parents can be reassured that their children are different from the moment they are born. In fact, even though they start with the same genes, single-egg twins are never identical for two main reasons: both their environments and their genes are different.

Consider their environment. No man is an island, and this also applies to single-egg twins. Our environment is a powerful influence on us all, and this starts long before we are born. Single-egg twins can have a different start in life because they implant at different places in the uterus, or use different regions of the same placenta. Because of this, they may have different degrees of access not only to food, but also to poisons and infections in their mother's bloodstream. Of course, their birth processes may be different as well, but all this is just the beginning – all through life, single-egg twins experience different things in different ways. We are, after all, not just the product of our genes.

But what about these genes? Surprisingly, single-egg twins are always genetically different as well. This may seem nonsensical – surely the whole point of single-egg twins is that they get the same genes? Indeed they do, but differences soon start to creep in, so that the genes of single-egg twins are rarely the same. As the cells in the newly separated twins multiply to form a whole person, errors in gene replication alter genes in different parts of their bodies. This is a subtle but very real effect, and causes some of the differences between single-egg twins, sometimes to the point where one twin has a particular genetic disease and the other does not. One disease where such a difference has been reported between single-egg twins is Beckwith Weidemann syndrome, a genetic condition that causes gigantism, hernias, tumours, heart abnormalities and an enlarged tongue and internal organs.

There are yet more ways in which genetic differences accumulate between single-egg twins. For example, genetic imprinting, in which genes are switched on or off depending on the parent from which they are inherited, is often carried out in a rather patchy way throughout the imprinted body tissues. This patchiness is fairly random and usually creates a different arrangement of imprinted and unimprinted tissue in each twin. Imprinting is therefore another factor that makes single-egg twins different.

Finally, there is another way in which female single-cell twins can become different. This is because girls 'switch off' certain genes in every cell in their body – a process called X inactivation. This gene inactivation takes place in an apparently random way in different parts of a girl's body and it lasts all her life. As many genes are switched off in this way, two single-cell girl twins can be very different depending on which bits of their bodies use which genes.

In one pair of single-egg twins, different patterns of X inactivation resulted in one sister being wheelchair-bound with muscular dystrophy, while her sister was a gifted athlete.

So 'identical' twins are not really identical at all. All these influences act on single-egg twins to drive them apart both physically and mentally. It is not just a moral or ethical judgement to say that single-cell twins are distinct individuals. The simple biological fact is that they are always different.

In around one pregnancy in every 100,000, identical twins fail to separate completely, producing what were once called 'Siamese twins' or even 'monstrosities', but are now called conjoined twins (Figure 3(d)). Conjoined twins are most commonly joined at the rump, belly or chest, but are more rarely connected at the top or back of the head. Occasionally, separation can be even less complete and the twins can share limbs, or even a torso and limbs. Very rarely, one twin is much smaller than the other: a 'parasitic' twin. Some forms of conjoined twinning can be separated surgically, although this is impossible for the many conjoined twins that share a single heart. The ethics of twin separation are very unclear, as twin separation can be an extremely risky business – often more risky for one twin than for the other. Surgeons are often faced with the dilemma of deciding which body parts should be given to which twin.

Among the most famous parasitic conjoined twins were the 'two-headed boy of Bengal', born to a poor Indian family in 1783. Although their midwife tried to kill them, the boys' parents realised that there was money to be made from their unfortunate sons, and the boys spent most of their lives being displayed for money. One of the twins was a normal size, but perched upside down on his head was a second inverted head. This second head ended in a stump-like appendage which presumably was a vestigial body. Perhaps mercifully, the exploited, and now emaciated, two-headed boy was killed by a snake bite at the age of four and buried. As if to extend their humiliation, their grave was plundered and the conjoined heads were brought to England for examination. Although the two skulls were fused, they contained two entirely separate brains. Had the twins been born today, surgeons may have been able to separate them, although of course the parasitic twin would not have survived.

Heads or tails, inside or outside?

After the embryo has hatched out of the zona pellucida, the inner cell mass starts to organise itself into something looking like a baby. This involves some contortion, folding and wrapping: a kind of embryonic origami. It can be difficult to visualise these changes, so hopefully Figure 2 will help.

By the fifth day after fertilisation, the embryo is a hollow ball of trophoblast cells with the inner cell mass inside (Figure 2(a)). For the next week, the embryo devotes most of its attention to burrowing into the wall of the uterus and growing enough trophoblast to produce hCG to stop its mother menstruating. The inner cell mass is also active during this week, however, reorganising itself into two hollow bubbles (Figure 2(b)). One, the early ectoderm ('outer skin'), is attached directly to the inner surface of the trophoblast. The other bubble, the early endoderm ('inner skin') dangles from the early ectoderm bubble into the middle of the blastocyst. Both of the bubbles are important, but the bit where the endoderm and ectoderm bubbles touch is the really exciting bit, because this is where the baby is going to be built.

The endoderm and ectoderm form a double layer where they meet. This double-layered structure is probably a relic from the time when our ancestors were made of two layers of cells, like jellyfish, hydras and corals today – all other many-celled animals, except sponges, are made of three layers. Around day 14, some ectoderm cells start to migrate to form the mesoderm ('middle skin') – a continuous sheet sandwiched between the endoderm and ectoderm (Figure 2(c), (d)).

Now, by day 17, the rudimentary baby is made up of three layers, the germ layers, forming a flat circular sandwich. This may seem an uninspiring start, but great change is soon to take place within this sandwich: the upper ectoderm slice is destined to form the baby's skin and the lower endoderm slice will form the lining of its gut. Most remarkably of all, the mesoderm filling will form almost everything else. So by the time it is two and a half weeks old, the embryo has neatly divided itself into three layers, each of which has a fixed destiny. Over the next ten days, the embryo will perform a feat of origami that will leave it looking recognisably like an animal, although maybe not yet a child.

As the origami gets under way, the embryo decides which way round it wants to be. A circular sandwich has no head or tail, but an embryo has, so something has to give it a sense of direction. An early sign of the orientation of the embryo is the formation of a rod-like structure in the mesoderm filling of the sandwich (Figure 2(e)). Discovered by von Baer, this rod is called the notochord and it establishes the orientation of the body of all vertebrates, as well as in some of their relatives.

Although it has a rather undistinguished final fate (it becomes the jelly-like bit that pops out when you slip a disc in your back), in the embryo the notochord is king. First of all, it dictates the position of the baby's back. In addition, the notochord also controls where the baby's mouth and bottom will be – and as any parent knows, these are the only bits that really do much for the first six months of life. The future head end of the baby is decided when a little patch (the mouth plate) appears at one end of the notochord – the head end. At this patch, a hole will eventually be punched through all three germ layers to make the mouth. Similarly, at the other end of the notochord a similar patch (the anal plate) is forming, and this is destined to become the embryo's anus.

Now that the three-layer embryo knows its head from its bottom, it is time to fold up into an animal-like shape. The three-layered sandwich now starts to curl down at the edges, rather like a real sandwich that has passed its prime. In the case of the embryo, however, this curling continues until the edges of the embryo meet underneath (Figure 2(e), (f), (g)). By day 28, the flat embryo, with ectoderm on top and endoderm underneath, has rolled up into a hollow ball with an ectoderm skin on the outside and an endoderm gut tube on the inside (Figure 2(g)). The space between the skin and the gut is filled by mesoderm, and within this the notochord runs from the front to the back of the embryo, above the gut. At the front of the gut is the mouth plate and at the back is the anal plate. Apart from having no legs, the embryo now suddenly looks animal-shaped (Figure 2(h)).

Not satisfied with this feat, the embryo has also managed to make the rest of its protective membranes. First of all, the original early endoderm bubble is left hanging off the embryo's belly at its navel (Figure 2(h)). This endoderm bubble is called the yolk sac, although human embryos do not actually keep any yolk in it. Behind the yolk sac is another, smaller bag, the allantois ('sausage-shaped'), and this will eventually form the bladder and the umbilical cord.

The folding of the embryo has stretched the early ectoderm bubble so that it now almost entirely surrounds the embryo. This bubble is now called the amnion (Greek for 'lamb'), and it contains the gelatinous amniotic fluid that will lubricate the baby's passage into the outside world in eight months' time. The amnion now forms the baby's inner membrane, within the outer trophoblast membrane (Figure 3(a)).

If things go wrong during all this origami, then the outlook for the embryo is usually pretty bad. Without a correctly arranged body, it will most likely become one of those pregnancies that fail almost as soon as they have started. However, very rarely, the folding of the body leads to a condition called fetus-in-fetu, which leads to the birth of healthy babies, but which must be one of the most unsettling medical conditions you can have.

A recent case of fetus-in-fetu was reported in Cairo in 1994. A woman had complained of pain during the sixth month of her pregnancy, and an ultrasound scan showed a strange mass within her baby boy's abdomen. When the child was born, his abdomen was mysteriously enlarged, and the strange symptoms were not explained until the boy was three months old. In a delicate operation, three fetuses were carefully removed from the outside of his large intestine and stomach. These 'inner siblings' do not always become apparent so early in life, and some people carry them into old age before they are discovered.

Although fetus-in-fetu may sound like an elaborate hoax, there is no doubt that it really happens, and we believe we know how. Unlike the Cairo case, there is usually only one fetus lodged inside its sibling host. The condition starts with two identical twins that share the same outer trophoblast membrane. It is thought that as one embryo rolls up into a ball, it somehow traps the other twin inside itself, so that embryo ends up within its abdomen, or less often its scrotum. The entrapped embryo then somehow manages to attach to one of its twin's organs, and draws enough sustenance to stay alive and grow. As we will see, the brain is another organ that forms by rolling up, and extremely rarely a fetus-in-fetu is found within the skull.

For a reason that I am unable to explain, the phrase 'foetus-in-foetu' also makes an appearance in one of the more impenetrable passages of James Joyce's *Ulysses*.

So, as long as there are no unwelcome lodgers, by day 28 the baby's body plan is established and the scene is set. All that remains is for the organs of the body to develop in their allocated places. In the rest of this chapter, we will follow the development of the four major organ systems that will make up the baby's body. First, we will see how the notochord causes the spinal cord to form along the back, as well as the bones and muscles that will support and move the baby's body. Second, the gut has to twist and contort its way into its final shape, also making lots of other organs in the process. Third, the nervous system, back and gut must cooperate to produce that ultimate constructional triumph, the head. Fourth, the formation of the kidneys leads to the creation of the all-important reproductive organs – already within the embryo the seed of the next generation is being sown.

Making the machine – brain, spine and limbs

In 1909, Charles Walcott, the secretary of the Smithsonian Institution, discovered the most important set of fossils in the history of science. Or rather, his horse tripped over them. Although he did not appreciate it during his lifetime, the strange imprints he found in the rocks 2500 metres up in the Rocky Mountains of British Columbia were to give zoologists a unique window into the oceans of 600 million years ago.

Although they are more than twice as ancient as the first

dinosaurs, the fossils of the Burgess shale look as well preserved as last year's pressed flowers. The aquatic animals that left these imprints died in a usually placid environment interrupted by catastrophe – they sank slowly to the bottom of the sea and were buried, possibly by mudslides, before they could become damaged. Because of their sudden burial, many Burgess shale fossils have preserved imprints of the animals' soft tissues, and not just their skeletons.

Despite the outstanding state of preservation of the fossils, their importance was completely underestimated until the late 1960s. Walcott himself studied them intensively, but he wrongly decided that the preserved animals were close, albeit ancient, relations of many of today's marine animals. Not until the work of Professor Harry Whittington at Cambridge University did their true nature become clear. His early papers on the Burgess animals in the 1970s showed that many of these animals are unlike anything alive today. They hark back to a time before today's familiar animals took over the seas.

At the time these fossils were laid down, the animal kingdom had not yet restricted itself to the central core of just a few body designs that we see today (vertebrates, starfish and their relatives, crustaceans, insects and spiders, molluscs, earthworms, round-worms, flatworms, corals and sponges). The diversity of animal bodies preserved in these shales is bewildering, and scientists cannot allocate many of the animals to modern groups. Indeed, it has even been difficult to work out what all the fossilised bits and pieces actually are and which way up some of the animals should be. Perhaps it is not surprising that one creature, which looked rather like a cheap alien costume from *Star Trek*, ended up with the name *Hallucigenia*.

Within all this diversity one animal, *Pikaia*, looks strangely familiar, and it gives us some clues to our own past. It was an eel-like creature with segmented muscles running along its body, just like present–day fish (think of smoked salmon). *Pikaia* and its kin were among the few Burgess animals to produce descendants that have survived from that era to the present day. We now think that the biological invention that gave the impetus to the success of

Pikaia and its relatives, the vertebrates, was the notochord. Because of this, all modern vertebrates have a notochord running along their back with a tubular gut suspended underneath. This is exactly the arrangement in the day 28 human embryo.

The notochord originally evolved in the larvae of our distant ancestors to help them swim through the primeval seas. Some of the closest relatives of the vertebrates, such as sea squirts, are filter-feeders stuck permanently to the sea bed. They feed by sucking sea water into a huge basket-like inlet, the pharynx, which sieves out food particles that are then digested in the gut. Although this sort of basket-animal may not seem very much like us dynamic vertebrates, the immature larval stage of the sea squirt is altogether more familiar.

Many sedentary sea animals have mobile larval stages to help them disperse through the environment, and the sea squirt is one such animal. The sea squirt larva already has the large filtering basket and gut of the adult. Behind and on top of this, it also has a rigid rod that it flicks from side to side with a series of swimming muscles. The whole animal looks not unlike a tadpole, and it swims in a similar way. This rigid rod is the larva's notochord, and this is why we think sea squirts are related to us. Unlike our notochord, however, the sea squirt's is temporary and it is discarded when the larva comes to rest on the sea bed and matures into an adult.

Vertebrates are like sea squirt larvae that never grew up – they keep their notochord all their life, and this is why they remain mobile. So the evolution of the vertebrate body may be an excellent example of Garstang's paedomorphosis: we have retained a larval characteristic of our ancestors into our own adult life. All this explains why vertebrates are constructed as they are: a mouth and a gut hanging off a rigid back.

The notochord may be sufficient to support something as diaphanous as a sea squirt tadpole, but it is not strong enough to hang a human body on. One of the notochord's jobs in human embryos is to induce the development of its own replacement, a bony back. The other structure that the notochord helps to build is the nervous system. This is why your spine not only holds you up, but also encloses and protects your spinal cord – all these structures are legacies of the notochord.

The nervous system is the first of the body's organ systems to start to develop. Around day 19, the notochord causes a change in the ectodermal 'skin' cells overlying it. These cells start to dip in to form a furrow that runs much of the length of the embryo's back (Figure 2(e)). This furrow deepens until it is a trench and eventually the two rims of the trench start to tilt towards each other. These two edges then meet and seal the trench off from the outside world, forming a tube (Figure 2(f), (g)) that runs along the entire length of the embryo's back by day 28. The ectoderm cells lining this hollow neural tube have a very different destiny from their fellows left behind on the skin. This tube is going to be the brain and spinal cord, and the ectoderm cells are going to become its nerve cells.

So all the amazing complexity of the human mind is derived from a simple buried tube of cells that is formed within a month of the sperm entering the egg. Different parts of the neural tube soon start to form the various parts of the nervous system. The spinal part of the tube changes the least, and in adult life the spinal cord is still a thick-walled tube with a canal down the middle.

The brain develops as four swellings at the front end of the neural tube, which start to grow around day 30. The hind-most swelling is called the hindbrain and it will carry out functions such as hearing, balance, coordinating movements, controlling the guts and heart as well as moving the muscles of the head. The second swelling is the midbrain which will eventually arrange information coming in from the eyes, act as conduit between the body and forebrain, and manage many of the hormonal processes of the body (the pituitary dangles from the midbrain). The other two swellings lie at the front of the neural tube, hanging off each side of the midbrain. These forebrain swellings are destined to grow much faster than the rest of the nervous system and will form the two huge cerebral hemispheres that dominate the brain by the time of birth. The forebrain carries out our more 'conscious' activities, such as speech, planning, memory, conscious movements and interpreting the inputs from the five senses.

The human forebrain's status as the location of our 'higher' thought processes belies its mundane evolutionary origins. The forebrain probably originally developed as a centre for interpreting smells — a function that it still carries out in humans. Nerves returning smell information from the nose have a uniquely direct route into the forebrain. The ability of the sense of smell to induce emotions, and even the feeling of *déjà vu*, has been claimed to be due to its unfettered access to the parts of our brain that carry out memory and emotion. By contrast, vision, hearing, taste and touch information has already been well analysed in the hindbrain and midbrain by the time it reaches our conscious brain, rather like a secretary passing on only the important mail to the boss.

Although it is the first organ to start to form, the brain is the last to be complete. The four main sections of the brain are established by day 35, but the growth and arrangement of all the brain's connections will not be complete until after birth. So the brain becomes progressively more complex throughout pregnancy, unlike most other organs, which are pretty much complete, albeit tiny, by the fiftieth day. This is one reason, other than the brain's inherent complexity, why damage to the baby at any time during pregnancy can affect the brain. All in all, the brain is rather exceptional, even problematic. Not only does it develop over a long period, rendering it especially vulnerable during fetal life, but soon after the baby is born its brain 'shuts up shop' as far as growth is concerned and becomes one of the few body tissues that cannot repair itself after injury. The tremendous advantages conferred by the mammalian brain come at a price — and that price is the fetal brain's vulnerability.

Because it forms as a tube, the baby's brain has a central canal that is continuous with the one in the spinal cord. This canal system remains intact into adult life and is continually flushed through with fluid. This cerebrospinal fluid is made in the forebrain and drains backwards into the spinal cord. At the end of the spinal cord it seeps out and drains away.

Unfortunately, the tubular design of the nervous system is the cause of some of the commonest birth defects seen in human babies. In about one baby in a thousand, the neural tube fails to close and the spinal cord and bony spine may be left open to the outside world, causing spina bifida. In its most severe forms, spina bifida can result in paralysis and learning difficulties. Also, if the tube fails to close at the head end, then the brain cannot form – a condition known as anencephaly. Anencephalic children are often stillborn, or die within their first few days. Occasionally, however, they can be supported for months or even years, although the ethics of this are controversial. Anencephaly usually has a genetic cause, but the mother's nutrition, especially insufficient folic acid in the diet, has also been implicated in the condition.

Even if the neural tube closes properly, the baby is still not out of danger. In perhaps one in a thousand pregnancies, blockage of the drainage of cerebrospinal fluid through the canals of the brain causes a build-up of pressure. This condition, hydrocephaly, is often an inherited genetic condition, but it can also be caused by infections. Under pressure, the canals swell and can squash the developing brain tissue. Severe hydrocephaly can damage large amounts of brain tissue before it is detected, but more minor forms can be treated by an operation to insert a tube which then drains the cerebrospinal fluid.

Spina bifida, anencephaly and hydrocephaly are some of the main reasons that maternity departments spend so much of their time examining babies in the womb by ultrasound. These abnormalities are relatively easy to diagnose by ultrasound, and spina bifida and hydrocephaly can somehow be treated even before birth.

While the brain is developing, structures are also forming in the embryo that will allow the brain to move the body. Most of the body's bones and muscles are produced as part of the notochord–back complex. Soon after the rudimentary nervous system has formed, it induces a change in the nearby mesoderm. All along the body, mesodermal cells form a series of bumps on either side of the neural tube. These clumps grow into discrete chunks called somites, spaced evenly along the embryo's back, rather like the crimped seam along the top of a Cornish pasty (Figure 2(i)). This even spacing of somites is the first sign that people are fundamentally 'segmented' animals.

Vertebrates, just like earthworms, are made up of lots of segments lined up in a row. The only difference is that the segmentation of earthworms is much more obvious from the outside. Despite this, segmentation is apparent in the internal anatomy of all vertebrates, including us. For example, the vertebrae in your spine and the stripes of muscle in a salmon are both products of the segmented somites. The somites make a very large amount of our bodies – they each divide into three pieces, which are destined to make skin, muscle and bone, respectively.

The skin blocks migrate away from the nervous system to lie beneath the skin ectoderm that covers the body. When they do so, they become the dermis, the lower layer of the skin. While the ectoderm produces the flaky cells that make up the actual surface of the skin (the epidermis), the dermis forms the tough connective tissue that makes the skin so robust, as well as the blood vessels that supply it with nutrients.

Each somite is connected to the spinal cord by its own segmental nerve, and because of this each skin block draws out some of this nerve to supply the sense of touch in the skin. As a result, the skin sensation of the human body is fed into the nervous system from a series of strips of skin corresponding to the areas formed by the respective skin blocks. A strange effect of somitic skin segmentation becomes apparent when people get shingles. Because the chicken-pox virus that causes the disease hides in segmental nerves, when it re-emerges in later life, shingles neatly picks out strips of skin that correspond to somitic skin blocks.

The cells of the muscle and bone blocks form most of the body's muscle and bone. Some of the cells stay where they are and form the segmented bony vertebrae around the spinal cord, as well as the muscles that move the back. (By the way, animals' tails are just a limbless, gutless extension of the back.) Another set of cells migrates into the developing limb buds to form the bones and muscles of the arms and legs. This is obviously a major undertaking, but it is completed relatively quickly – the paddle-shaped limb buds first appear at day 26, and the arms and legs are pretty well mapped out by day 50.

Embryologists have made great efforts to find out how limbs form. Few parents can have failed to wonder at the perfection of their newborn child's tiny fingers and toes. It is remarkable enough that a baby can be produced at all, never mind that it has exactly five digits on each hand and foot. The construction of an arm or leg is coordinated by the apical ectodermal ridge – a ridge of ectoderm at the end of each limb bud. As the limb grows, the ridge constantly communicates with the mesoderm in the limb by an array of chemical signals. As the limb lengthens, it is constructed 'from the torso outwards' – for example, the thighbone is generated before the bones of the calf.

Once the main parts of the limb have been established by day 35, the hands and feet form. The tissue between the future fingers thins and eventually disappears by day 50, turning the flat limb paddle into a five-fingered hand or a five-toed foot. By halfway through pregnancy, the fetus is able to clench its fist, much to the irritation of ultrasonographers trying to count its fingers. The thinning of the tissue between the fingers is actually caused by cells dying. In recent years, biologists have discovered that many of the processes that form and maintain the human body are based on a special kind of cell suicide called apoptosis. For the hand to form normally, chemical messages instruct the cells between the future fingers to self-destruct – to sacrifice themselves for the good of the rest of baby. So a new life is produced by a continual process of death.

Occasionally, the subdivision of the limb paddles into digits goes awry and babies are born with extra digits (polydactyly) or fused digits (syndactyly). Polydactyly, of which one of the most famous alleged cases was Anne Boleyn, is usually an inherited genetic condition. It is especially common in cats, and it is often more common in some areas than others. Polydactyly also raises intriguing questions about the evolution of the hand and foot. Until recently, it was thought that no normal land vertebrate could have more than five digits per limb. Many animals have reduced the number of digits to suit their needs – horses have one toe per limb, birds have four toes per foot, and frogs often only have four fingers – but no normal living vertebrate has more than five digits. In recent years, palaeontologists have discovered fossils of some of the very first vertebrates to leave the water and take their chance on dry land. These creatures had well-developed digits, but they often had six, seven or eight of them. These supernumerary digits seem to spill off the end of the hand and continue along one side of the forearm – exactly as they do in polydactyl cats. So it seems that at some stage following our ancestors' emergence on to dry land, a developmental rule was introduced that restricted all land vertebrates to just five digits. Presumably the gene that enforces this rule is damaged in animals with polydactyly.

The first time that a mother knows that her baby's nervous system, skeleton and muscles are working is when she first feels it move. For many women, this can be the first time that they truly believe they are pregnant. The fetus is usually first felt to move in the fifth month of pregnancy, but for some reason it can often be felt earlier in women who have already had a child. Many women describe early fetal movements as feeling like a little butterfly trapped in their stomach, but within a couple of months babies often become altogether more violent.

Ultrasound scans show that early embryos are extremely active – they often perform somersaults. However, fetal movements are extremely erratic, and fetuses often seem to stop moving for a few days simply to panic their mothers. Many mothers learn tricks to make their babies move on cue if they are worried. The combined caffeine and sugar 'hit' of chocolate is often a good way to rouse a fetus into action, and others may respond to particular voices. Mothers can also become worried if their babies feel like they are hiccuping or even having a fit. Yet no matter how outlandish a fetus's movements are, they are almost never a sign that anything is wrong. Also, in the last month of pregnancy, fetuses often start to move less because they are running out of kicking space by this time.

Churning guts

When we last looked at the gut at around day 20 of embryonic life, it was just a hollow endoderm-lined tube going straight from the head end of the embryo to the anus (Figure 4a). Over the next few weeks, this simple tube will contort into the final complex shape of the human digestive system (Figure 4b).

Figure 4. The contortions made by the developing gut to produce the lungs and intestines. These schematic front views of the chest and abdomen of a developing baby show how the straight gut tube twists to produce the stomach and intestines, and also how it develops pouches that form the lungs, liver and pancreas.

(a) Day 21   **(b) Birth**

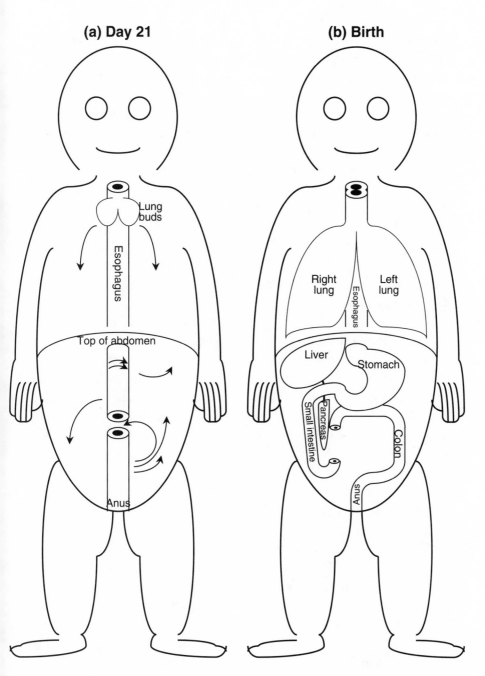

The section of the gut that lies in the chest is the gullet or esophagus. This part of the gut is quite simple, and it keeps the straight tubular structure of the original gut even in adults. As with all the other regions of the gut, endoderm forms the cells that line the inside, and surrounding mesoderm forms the muscles that allow the esophagus to squeeze food down into the stomach.

The esophagus is not quite that simple, however. At about day 21 a very important pouch branches off the front end of the esophagus. This endoderm-lined sac grows in the direction of the belly until it splits into two smaller pouches, one on the left and one on the right (Figure 4(a)). These keep on growing belly-ward until they in turn split into several branches. These branches then branch repeatedly until two complex trees of tiny blind-ending pouches are hanging off the esophagus into the chest. These develop a profuse blood supply and will eventually become the largest organs in the chest, the lungs. Like the rest of the organs made by the gut, the lungs will not really get used until birth, although they must be ready immediately when birth comes.

Beyond the esophagus, all the rest of the gut is in the abdomen. Really, we should not call it the abdomen yet, because it is still not separated from the chest. The abdomen and chest are separated by the diaphragm, a curtain-like muscle that develops from a bar of mesoderm in front of the navel between days 35 and 50. This bar grows up into the body cavity until it forms a continuous sheet of muscle that divides the chest from the abdomen. The diaphragm is not just a way of separating chest from abdomen, however. When the baby is born, it will be the most important muscle for breathing.

The abdominal portion of the gut is, of course, the 'business end' during later life. In a series of contortions, the abdominal gut arranges itself into its final configuration by day 50. The twisting of the gut is thought to be caused by changes in the rate of growth of different bits of the gut tube. For example, the gut tube as a whole grows much faster than the embryo in general, and so there is a lot of 'slack' which piles up into coils of intestine. (The twists and turns that the straight gut tube makes to reach its final state are depicted in Figure 4. I have included only the first and last thirds of the abdominal gut in the picture because these are the two portions that become fixed in adult life.)

The front part of the abdominal gut becomes extremely broad and muscular, and most of it flops over to the baby's left-hand side. This is the stomach and it will become tightly fixed in this position to stop it flopping around and getting twisted as it fills and empties during later life. Because of its contortions, the stomach now passes from left to right and empties into the duodenum on the right-hand side. The duodenum is the first part of the small intestine and it passes down the right-hand side of the abdomen before turning left and head-wards again.

Two vital organs grow out of the duodenum. Like the lungs budding from the esophagus, two blind-ending sacs now sprout out of the duodenum near the stomach. Like the lungs, these grow and branch into complex trees of ducts draining into the gut. The smaller bud grows tail-wards into the tissue holding the duodenum and stomach. It will eventually make the hormone insulin and secrete enzymes to help to digest food: this is the pancreas. The larger bud grows head-wards into part of the bar that formed the diaphragm. This bud will become a large rubbery organ that carries out a multitude of useful biochemical reactions, as well as secreting bile into the gut to digest fats. This is the liver and it will come to lie on the right-hand side of the abdomen in humans, jammed up against the diaphragm.

The rest of the small intestine grows to fill the front part of the abdomen and will absorb most nutrients from the baby's food. The small intestine develops mainly by a spectacular elongation that produces a prodigious mass of writhing coils. The small intestine is more mobile than the rest of the gut and churns around the abdomen during normal digestion. This is why I left it out of the diagrams – it does not really have a 'normal position' at all. The small intestine is also the part of the gut that remains attached to the body stalk and yolk sac at the navel. Because of its tremendous mobility, some of the small intestine protrudes from the navel and develops outside the abdomen for a while. This umbilical hernia usually, but not always, falls back into the abdomen before birth. Umbilical hernia is a common condition and ranges from mild to severe. Sometimes there is only a tiny hole at the navel and only a small blob of fat can fall through. In other cases, the hole is large and several loops of intestine can fall through. In fact, the most dangerous form of umbilical hernia occurs when the hole is medium sized – these holes may be large enough to let intestine out, but not large enough to let it back in again. If the trapped intestine twists and cuts off its own blood supply, then the damaged gut may have to be removed and the baby's life can be threatened. Fortunately, most cases of umbilical hernia can be treated by relatively simple surgery.

The back end of the abdominal gut, like the front end, undergoes a dramatic twisting movement during development, but becomes fixed in later life. The gut tube edges its way along the left side of the body and then curls around under the stomach to connect to the end of the small intestine on the right-hand side. This is why the colon and rectum end up shaped like a question mark (Figure 4(b)). The anus forms when the anal plate breaks down so that the rectum opens on to the outside world. Failure of the anal plate to break down completely is a relatively common congenital defect, but provided that the rectum reaches as far as the anal plate, correcting what is known as an imperforate anus is quite an easy operation.

So here we are, at the end of the baby's intestines. The different parts of the gut have made their tortuous journeys to their final destinations by around day 50. For the rest of pregnancy the gut will be relatively quiescent, although later on in pregnancy the fetus starts to swallow amniotic fluid. The gut has a trial run at digestion by converting this fluid into meconium – tacky green fetal feces which is stored up in the rectum to be graciously bestowed upon the baby's first few nappies. Rather like cutting a ribbon in front of a supermarket, the appearance of meconium heralds that a baby's gut is open for business.

How to get a head

You will have noticed that I have carefully ignored the parts of the gut that lie in front of the esophagus. You may also have realised that I skipped over all the connections that the brain makes with the head. But now we can postpone the inevitable no longer: it is time for the baby to assemble its head, a complex jumble of eyes, ears, nose, muscles, skull, jaw, throat, tongue, glands, blood vessels and nerves, not to mention the brain. Fortunately, when embryologists first discovered how the head and neck form, this apparently chaotic mixture of unrelated bits and pieces suddenly seemed to make sense. The easiest way to understand how the head is put together is to think of it as a modified version of a much simpler animal – a sea squirt larva.

The sea squirt larva is the bag-of-guts-on-a-stick creature that I mentioned before. It has a basket-like pharynx that it uses to filter its food from seawater, and it is propelled forward by a segmented muscular tail at the back. The sea squirt pharynx is equivalent to the throat of a baby. The sea squirt's segmented tail is equivalent to the somites running along the back of a vertebrate embryo, and these somites are also present in the head. Because the head has incorporated both the pharynx and the tail of our sea squirt-like ancestors, it has two separate patterns of segmentation – somites on top and gill bars in the throat (Figure 2(i)).

As well as the somites and gill bars, there are two other, unsegmented systems that produce important structures in the head: the sensory plaques and the dermal armour. The sensory plaques are thickenings of the skin of the outside of the embryo's head and these eventually form the ears, eyes and odour-sensitive parts of the nose. The bony armour is descended from the large bony plates that covered the heads of our distant vertebrate ancestors, and it is now used to make large sections of the skull.

Let us start at the top and back of the head and work forward and down through the sensory plaques, the somites, the bony armour and the gill bars.

Head part 1: the empire of the senses

By day 28, three pairs of sensory plaques can be seen on the top of the baby's head in roughly the positions of the future nostrils, eyes and ears (Figure 2(i)). Each plaque is a thickening in the ectodermal 'skin' that will form one sense organ.

The nose plaques sit just above the primordial mouth. As they develop into smell-sensitive membranes, they also cause part of the forebrain to bulge out to meet them. This connection will form the olfactory tract, an outgrowth of the brain which passes smell information back to be interpreted by the brain. By day 40 the smell-sensitive membranes have become buried deep inside the front of the head by the nose that forms around them. Folds of tissue grow forward to produce the top and sides of the nose, as well as much of the face. Other folds then grow between the plaques to divide the two nostrils, and below the plaques to form the palate that separates the nose from the mouth. Failure of the palate to form completely is one of the commonest congenital defects in children, but hare lip and cleft palate can now usually be repaired surgically.

The eye plaques, of course, have a rather exotic future. These plaques sink into the head to form bubbles of ectoderm stranded in mesoderm. These bubbles are destined to become the lenses of the eyes. Meanwhile, stalks have grown out from the brain, and these soon expand to produce structures shaped like brandy glasses. A lens lies suspended across the mouth of each brandy glass, and the glass itself now becomes most of the rest of the eyeball. Much of the brandy glass becomes the light-sensitive retina, but the lip of the brandy glass becomes a circular ring of coloured muscle, the iris, that partially covers the lens. The stem of the glass becomes the optic tract, and this soon starts to carry visual information from the retina to the brain.

Just a few finishing touches are needed and the eye will be complete. Skin overlying the lens forms the glassy front surface of the eye as well as the eyelids. The lids are sealed closed at first but they are open later in pregnancy, just as they are in newborn babies (well, baby humans anyway – puppies and kittens have to wait until they are a few weeks old).

The ear plaques form at day 21, further back on the head. Like the eye plaques, the ear plaques sink into the side of the head to form hollow bubbles. The ear bubble now undergoes a series of dramatic contortions to form the 'labyrinth' of the inner ear by day 50. The labyrinth, as its name suggests, is a complex hollow structure, but, put simply, it is divided into two sections that carry out the two different jobs of the inner ear: hearing and balance. The lower part elongates into a long tube, which coils up so that it looks like a snail shell. This is the cochlea, the organ that detects sound vibrations. Inside the cochlea, a series of delicate sheets resonates with sounds of different frequencies – one end resonates with low-pitched sounds and the other with high-pitched sounds. The other, upper half of the labyrinth becomes even more complex, forming three hollow semicircular tubes, each arranged at right angles to the others. These tubes are destined to be the baby's most important way of controlling its balance. After the baby is born, it can detect movements of its head from the movements of the fluid inside these three tiny tubes.

There is one more sensory plaque in the head: the taste plaque. By a quirk of evolution, this structure gives rise to taste buds in people and to the lateral line in fish. The lateral line is the visible 'seam' passing all the way along the flank of fishes. The lateral line is extremely sensitive and can detect chemicals and electric fields in the water around fish. In people the same plaque migrates to the top surface of the tongue to form the taste buds, which are also specialised for detecting chemicals.

For some time, scientists have debated how aware fetuses are of their environments. Babies are obviously in a high state of awareness immediately after they are born, and it seems unlikely that this awareness is suddenly switched on during the birth process. So what do they experience while still in the womb? Babies are unlikely to see very much before they are born, but that is mainly because the uterus is so dark. It is likely that babies can see something while still within the womb, however, although we do not know what they see. Some light probably filters through to the baby's eyes, at least when its mother exposes her titanic midriff.

Fetuses' sense of touch is well developed almost as soon as they can be felt to move. Babies often react when their mothers adopt a posture that squashes the uterus, and many babies object to being prodded through their mother's belly by ultrasound probes. The question of fetal touch sensation is an important one now that surgeons are able to carry out operations on fetuses before they are born. At what stage can a fetus first feel pain? Studying pain is difficult, especially in something as precious as a human fetus. Until recently, even newborn babies were assumed to be relatively insensitive to pain. Because anaesthetists worried about the risks of anaesthesia in newborn babies, major operations were often carried out on babies who were effectively paralysed, and not anaesthetised. Despite the assumption that babies do not really feel pain, the levels of stress hormones in these babies were found to be alarmingly high, and this has led to a change in attitude to neonatal surgery. As a vet, I know that lack of response to pain does not imply a lack of pain – many animal species are notoriously poor at telling you when they are in pain.

The fetus' ability to hear has attracted a great deal of attention, partly because it seems the most likely way in which the fetus can be aware of the outside world. The uterus appears to be a fairly quiet environment, but the level of ambient noise is certainly well above the level that a mature ear can hear. Although a fetus' middle and outer ear are full of fluid, this does not mean that the cochlea cannot detect sound – after all, this is exactly how fish hear. As most pregnant women will attest, there is good evidence that fetuses can hear the outside world and can even respond differently to different voices. Because of this, many people believe that speaking to a fetus gives it its first lesson in the musicality of language. This may seem a rather speculative idea, and certainly it would be difficult to prove, but it is not such an unreasonable suggestion when you consider that most of the information that reaches the fetus from the outside world comes in the form of sound. The violinist Yehudi Menuhin even claimed that his ear for music was instilled before birth, as his parents spent so much of their time singing and playing musical instruments.

Knowing that fetuses can hear and distinguish sounds has led many parents to wonder if they can help their child develop by exposing it to certain sounds. Certainly, babies are thought to remember sounds they have heard in the womb: for example, their heart rates rise when they hear a tune they became familiar with before birth. Coupled with recent evidence that listening to well-structured but non-repetitive music can improve adults' reasoning skills, some parents have tried to give their babies a head start by exposing them to Mozart. Perhaps the best thing is for both parents to sing to their unborn child, as songs have the advantage that they contain elements of both music and language. Perhaps this explains why it is traditional in many cultures for women to sing to their unborn babies.

Head part 2: piece by piece – the somites

The second bit of the head is made of segmented somites. The head somites are a continuation of the body somites and extend right to the front of the head. There are probably eight pairs of them, but some of them are rather distorted, and some are obliterated by the developing brain.

The first three muscle blocks at the front of the head form the tiny muscles that move the baby's eyeballs. This is not as simple as it sounds: keeping your eyes fixed on an object as your head moves is actually quite a complex task, and it requires a great deal of discussion between the eyeball muscles and the brain. Because of this, of all the muscles in the body, these ones are capable of the most precise movements. The eyeball muscles are unusual in other ways that suggest that they may not be very like the other somites. They form much later than the other somites and they do not lie alongside the notochord.

The fourth muscle block is not really there – it is obliterated by the developing brain – and the same fate has befallen most of the fifth block as well. Behind the brain, muscle blocks six, seven and eight look a lot more like the trunk somites that lie immediately behind them. In fact, these muscle blocks may represent part of the ancestral fishy neck that has been drawn into the head at some point in our evolutionary history. Blocks five to eight all have the same fate: they migrate to the floor of the throat, where they become the intricate muscles of the tongue.

The somite bone blocks form part of the developing skull. The skull is like the medieval cathedral of the human body: an impressively complex structure of interconnected arches, windows and flying buttresses. However, the bone blocks just give rise to the cathedral floor – they form the base and back of the brain-box.

Head part 3: bony armour

The rest of the baby's skull is made by a different set of structures. This third group of head structures is the bony armour, and it has a very long history.

Around 350 million years ago, the first amphibians crawled on to land. They were not frogs and newts as we know them today. In fact, most of them looked a lot like little tanks. Many were newt-shaped, but instead of having a thin skin they were covered by large bony plates to protect the fleshy animal underneath. This bony armour is not made from somites, but can form spontaneously under the skin wherever it is needed. Originally, this bony armour lay on the surface of the head and protected the brain-box underneath. Yet vertebrate evolution has progressively fused this armour with the old cranium to make a unified, but dual-origin skull. In humans the armour has replaced much of the old brain-box and almost all of the old jaws. As a result, bony armour builds the rest of the cathedral: the barrel-vaulted roof that seals in the brain, the delicate tracery of tiny struts and beams that support the eyeballs, the light yet strong promontory of the muzzle and the crushingly powerful jaws. The skull is now the only relic of this ancient armour.

The other structures produced by the bony armour are a unique vertebrate invention: the teeth. Although there are signs of developing teeth in the embryonic jaws by day 40, tooth construction is a very leisurely affair, and of course the first set of teeth is not complete for some years. The first teeth to appear are usually the middle lower incisors, which often erupt around six months of age. By the time a baby is born, however, its jaws contain the rudiments of both sets of teeth. Occasionally children are born with a couple of milk teeth already visible, and very rarely they may have a full mouth of teeth. Extremely rarely, humans can produce three or even four sets of teeth during their lifetime.

Mammals are unique among vertebrates in that most of them

have only two sets of teeth during their lifetime – although some whales have only one set of teeth. Most other toothed vertebrates are much more flexible: they simply grow new teeth when old ones fall out. Because of this, they do not really have discrete sets of teeth as such – to them teeth are disposable items. Probably, the reason why mammals have restricted themselves to only two sets of teeth is that each set is just that: a set. Each tooth is designed to work faultlessly with all the other teeth around it. Unlike reptile and fish jaws, which are often full of mismatched fangs at different stages of development, the mammalian jaw is a precision tool. Mammals have one great skill, and that is chewing – in fact, chewing may well be the one thing that has got us where we are today, since chewed food gives up its nutrients more easily. Because of this, teeth have an importance far greater than their physical size would suggest and mammals, including humans, have restricted themselves to two sets of perfectly intermeshing teeth, rather than innumerable jumbled disposable fangs.

Head part 4: the throaty bits – the gill bars

The fourth and final set of structures in the embryo's head is formed by the gill bars. The seven gill bars of the human embryo are equivalent to the struts separating the gills of modern fish, but unlike fish, the gaps between them do not perforate out of the side of the neck to form true gills. Instead, our human embryo's gill bars lie along each side of the pharynx, separated by six shallow grooves.

Contrary to Haeckel's claims, these gill bars only become true gills in fish. Bony fish like cod and trout have discreetly covered over their gills with a flap, but you can still clearly see a series of five slits on the side of the throats of gristly fish such as sharks and dogfish. Modern fish breathe by drawing water in through the mouth and expelling it through these perforations in the sides of their pharynx. Because of this, each gill bar has its own artery to carry blood past the gills for oxygenation, before it courses backwards to take the oxygen to the rest of the body.

Haeckel's big mistake was to claim that human embryos have

actual gills early on in their development. This, of course, supported his idea that we pass through all our evolutionary history as we develop in our mother's womb. We now know that early human and fish embryos both share a pattern of pharynx segmentation, and in fish this segmentation develops into gills. Instead, the human embryo's gill bars and the grooves that separate them are used to make a wide variety of different structures in the human head and neck. The mammalian adaptation of the gill bars is a good example of pragmatic salvage and redesign of old structures that have outlasted their original function.

At the front of the throat is the first gill bar, and in most vertebrates this forms the upper and lower jaws. Mammals have dispensed with the gill bar jaw, however, and have replaced it with one made from bony armour. Yet nothing goes to waste in the embryonic salvage yard and the old jaw has been recycled as two of the three tiny bones in the middle ear. Once it is born, the baby's malleus and incus ('hammer' and 'anvil') will transmit sound from the eardrum through the middle ear to the inner ear. Because of this strange arrangement, mammals have two sorts of jaw muscle – one that opens and closes the jaws, and a modified jaw muscle that tugs on the tiny malleus to tauten the eardrum. This movement protects the ear from damage by loud noises.

Between the first and second bars, there is narrowing in the pharynx wall called the first gill groove. This breaks open in a fish embryo to make a gill, but in a human embryo it forms the 'pipework' of the baby's ears. The groove on the outside surface of the embryo deepens to form a tunnel diving far into the side of the head. This tube becomes the outer ear canal, and stretches from the side of the head to the eardrum. A corresponding groove in the throat tunnels outwards to meet up with the external tunnel. This inner tunnel is the Eustachian tube, and when it reaches the tiny ear bones, it envelops them in a liquid-filled cavity, the middle ear. After birth, this liquid drains down the Eustachian tube into the throat. The eardrum membrane forms where the outer and inner tunnels meet. The eardrum vibrates when sound is transmitted down the external ear canal and it transmits this vibration across the ear bones to the cochlea.

An exceptionally ancient structure is forming on the floor of the throat next to the first gill bars: this is the endostyle. In the middle of what will eventually be the tongue, a pit forms. In our filter-feeding distant ancestors, this glandular pit secreted iodine-rich mucus to bind up the food particles gathered by the ever-filtering pharynx. Modern vertebrates no longer need iodine-mucus to feed, so instead, the gland burrows through the tongue and migrates halfway down the neck. This gland will be our baby's thyroid gland and it now secretes iodine-rich thyroid hormones that control the body's metabolism. Occasionally the thyroid fails to form – a condition known as cretinism. Because thyroid hormones can cross the placenta from the mother, cretinism is not always apparent at birth. Once they grow up, however, cretins usually become short and mentally retarded, although cretinism is now easily managed with artificial thyroid hormone supplements. Although the word 'cretin' is now used as a term of abuse, it has a charming, indeed rather noble origin. It is a modified version of the old Swiss French word for 'Christian' – cretins were given their name to emphasise that, despite their problems, they deserved the respect due to all Christian souls.

The rest of the gill bars produce structures involved in swallowing and breathing. The second and third gill bars form the hyoid apparatus, which pulls the esophagus up to swallow food. This swallowing action also helps to close off the voice box or larynx, so that you do not inhale food. The larynx itself is made from the fourth, fifth and sixth gill bars. It has several jobs, including speech, coughing and stopping us inhaling food.

The second bar also has two other functions. First, it forms the stapes ('stirrup') ear bone, the smallest bone in the body. Second, the muscles that move the face are almost all derived from this gill bar. Unlike other vertebrates, mammals' faces are very expressive, and this is because of these facial muscles. So our baby's little face is soon animated by myriad tiny muscles swarming upwards from the throat to a more exciting life on the face.

The fourth gill bars also produce another major structure, the aorta. Each gill bar has its own set of blood vessels. In humans, the left fourth gill bar artery grows out of all proportion to the others until most of the blood coming forward from the heart goes through it before coursing down towards the trunk. Because of this, the fourth gill bar artery becomes the main artery of the body, the aorta, and this is why the aorta comes out of the top of the heart and swings over to the left before running down into the abdomen.

The rest of the gill grooves produce several more important structures. The second groove forms the palatine tonsil – part of the immune system's first line of defence against foreign invaders. The third and fourth grooves burrow into the neck and form a parathyroid gland each. These tiny rice grain-like glands remain lodged in the neck and start to secrete a hormone that maintains levels of calcium in the blood – these glands are essential because if your blood calcium levels fall below a certain level, you will die. The third and fourth grooves also make a structure called the thymus, which sets off on a long journey down into the chest where it comes to lie in front of the heart. As we will see in the next chapter, the thymus is an important part of the immune system. The fifth gill grooves make the C-cells, which migrate into the thyroid gland. These produce a hormone called calcitonin which counteracts the parathyroid gland by reducing levels of calcium in the blood – high levels of calcium are almost as dangerous as low ones.

So this is how the embryo makes its head, from the sensory plaques on top to the rack of gill bars below. The human head is the result of hundreds of millions of years of evolution; a custom modified version of a sea squirt's tail, an ancient amphibian's armour and a fish's throat. With so much heritage mixed up in the human head, perhaps it is not surprising that Haeckel thought that evolution and embryonic development were the same thing.

The baby has one more thing growing in its neck – something rather unexpected. The neck may not seem a likely place to find a kidney, but this is where the story of the embryo's kidneys and reproductive organs will start.

Down the tubes – planning the kidneys and the grandchildren

We need our kidneys in order to survive, but there is much more to them than that. The development of an embryo's kidneys is wrapped up with its eventual ability to produce its own children. Embryos' ovaries or testicles grow next to, and in intimate association with, their kidneys. They even incorporate unused pieces of kidney into themselves. By the end of the embryo's crucial first fifty days, it will have decided which sex it is going to be, and will even have started making its own eggs or sperm-generating cells.

The development of the kidneys is a part of human embryonic development that can rather perversely be explained largely in terms of Haeckel's now discredited biogenetic law. Being choosy little creatures, babies get through no fewer than two sets of kidneys before they finally get the ones they want, and this strange succession of kidneys parallels the evolution of the mammalian kidney. This is why the embryo has a kidney in the back of its neck. This first-kidney seems to be very much a relic from our ancestors and it is not thought actually to do anything at all. Like all later kidneys, the first-kidneys form in the mesoderm on either side of the attachment of the gut (the position marked 'X' in Figure 2(g)). We have to search quite far in our extended vertebrate family to find a first-kidney that actually does anything: in the jawless lamprey, the first-kidney drains fluid from the cavity surrounding the heart. The human first-kidney is rather ephemeral and redundant – its front end fades away even before its back end has formed, during the embryo's third week.

The second-kidneys form in the fourth week and are much larger. They appear in the same position relative to the gut as the first-kidney, but lie further down in the baby's chest and abdomen. The second-kidneys are functional and actually remove waste from the blood for much of the embryo's second month. The aorta sends off branches into the second-kidneys which filter waste into little urine-collecting ducts. The embryo has one of these Wolffian ducts on each side, and they carry urine from the kidneys to the developing bladder. If the embryo is male, the ducts will be salvaged and pressed into service as part of the male reproductive system. In boys, these ducts produce the sperm-storing tubes in the testicles and the vas deferens which carries sperm from the testicle to the penis for ejaculation. As you can see, the embryo is planning ahead.

Around day 34, a second pair of tubes forms alongside the second-kidney ducts. These tubes open into the abdominal cavity at their upper end and run downwards until they lie alongside each other, near the anus. If the embryo is female, these Müllerian ducts will join together to form the uterus. At this early stage, however, boy and girl babies' internal anatomies look identical and the Wolffian and Müllerian ducts are still present in both sexes.

Last in the brief kidney dynasty, the final-kidney forms down in the pelvis, also around day 34. Tubules grow out of the Wolffian ducts into the final-kidney and start to drain urine from it. This kidney is active, filtering out waste and poisons from the blood-stream, although the urine it produces has a rather tortuous journey before the embryo is completely rid of it. The embryo excretes urine into the amniotic fluid and from here it is swallowed. This waste is then reabsorbed from the gut into the bloodstream and eventually makes its final exit across the placenta. From then on, the embryo's mother has to deal with it.

A couple of weeks ago, the allantois formed as a bag connected to the gut, and now part of it hangs out of the navel. The section of the allantois outside the embryo's abdomen has been busy forming the umbilical cord, and from day 35 the portion left inside the embryo starts to become the bladder. All the kidney ducts connect up to this part of the allantois, and this is why urine from the kidneys drains into the bladder.

While all this construction work on the kidneys has been going on, the baby's ovaries or testicles have been making their first tentative appearance. They form in the same ridge of mesoderm as the kidney, slightly to one side of the second-kidney. A meshwork of fibres forms inside these organs, eagerly awaiting the arrival of the cells that will generate the eggs or sperm. For some unknown reason, these primordial cells originate outside the embryo in the yolk sac and embark on an extravagantly long journey. They start their life as mobile amoeba-like cells and they crawl up the yolk sac, through the navel, past the gut to the back and finally turn sideways into the developing ovaries or testicles.

At this stage, the embryo has 'indeterminate' ovaries/testicles: to look at, it could be either male or female. Yet the baby has had its sex pre-programmed by a simple, efficient mechanism ever since fertilisation. In the coming weeks, this programme will mould the sexuality of the baby. For those of you who have grown rather attached to the embryo that we have been following throughout this book, I can now reveal that it is a little girl.

Animals use many different methods of deciding what sex their children should be, but humans use a simple genetic switch provided by the sperm. The genes in every human cell are packaged into forty-six chromosomes: twenty-three from each parent. One of these twenty-three pairs is exceptional because the pair look alike in women (they are both X shaped) but in men one of the pair is a stunted Y shape. Because women almost always have an XX pair of these chromosomes and men usually have XY, biologists believe that having a Y chromosome makes you a male. These chromosomes are called the sex chromosomes.

Your sex was determined entirely by the sex chromosome you inherited from your father. Your mother is a woman, so she must be XX. Because of this, you must have inherited an X chromosome from her. However, your father was an XY man, so you could have got either an X chromosome or a Y chromosome from him. Your father bequeathed just one of these chromosomes to you, and he did this by randomly packaging X chromosomes into half of his sperm and putting Y chromosomes into the other half. If an X sperm reached the egg first, then you are female and if a Y sperm won the race, you are male.

We do not know for sure why we use this XX-female and XY-male method, or how it evolved in the first place. Certainly, there are other ways to use genes to control your offspring's sex. Birds, for example, are the other way round – cocks are AA and hens are AB, and having a B chromosome makes a bird female. One reason why we might use our system is that it produces roughly equal numbers of male and female children because X and Y sperm have similar swimming abilities. X and Y sperm are not absolutely identical swimmers, however, and doctors have used their subtle differences to develop artificial methods of separating them. Prospective parents keen to pay large amounts of money to use these techniques should be warned, however, that X and Y sperm separation is still a very inexact science.

We now think that our unusual X and Y sex chromosome pair probably evolved from a 'normal' pair of chromosomes. Most pairs of chromosomes swap genetic material with each other every time a sperm or egg is made, but the first step in the evolution of the X and Y chromosomes probably came when they stopped swapping portions with each other. Once the two chromosomes stopped mixing their genes they became distinct entities, able to evolve in different ways. As time passed the X chromosome kept most of the useful genes, while the Y chromosome threw most of them away, becoming the tiny stunted thing it is now. The one critical gene that the Y chromosome retained was the gene that makes embryos male.

The way our sex is controlled by our X and Y chromosomes has an important consequence for women. Because half the human population are XY, human cells have been designed to function with just one X chromosome. This has left women with a problem because they have two X chromosomes in each cell, leaving them vulnerable to an 'overdose' of the products of the genes on this chromosome. To get around this problem, female embryos switch off one X chromosome at random in every cell in their body and this X inactivation persists all through a woman's life. Each cell in her body has only one active X chromosome, just like a cell in a man's body.

X inactivation has interesting effects on the female body. Because of the female embryo's random choice of which X chromosome to switch off, some patches of an adult woman's body end up using one X chromosome and other patches use the other one. Because of this, every woman's body has two different populations of cells, each using different sets of genetic instructions – one reason why identical girl twins are never truly identical.

Soon after his discovery of the circulation of the blood, William Harvey
became personal physician to Charles I in 1630, and he frequently accompanied the king on hunting trips. Here Harvey demonstrates the beating of the
fetal heart of a deer to the king. Engraving by H. Lemon, 1851, after an oil
painting by R. Hannah, 1848. Courtesy of the Wellcome Library, London.

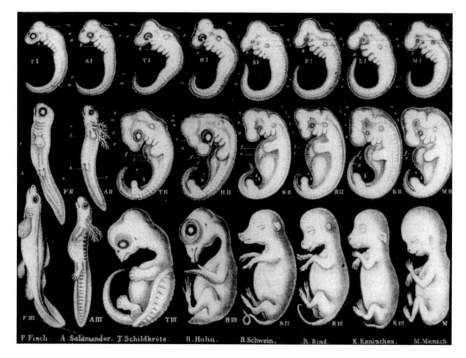

Ernst Haeckel's studies of animal embryos, as well as his own peculiar philosophy, led him to propose his controversial 'Biogenetic Law'. According to this 'law', embryonic development and the evolution of new species are essentially equivalent processes. Because of this, humans are claimed to pass through all stages of their evolution as they develop in their mother's womb, an idea that has proved remarkably influential. Haeckel used this image, showing how early embryos of different species look surprisingly similar, as evidence to support his 'law'. However, his ideas have gradually been refuted over the last century, not least because elements of this picture have been shown to have been faked. From Ernst Haeckel, *Erklärung von Tafel VI und VII.*

Ernst Haeckel was obsessed by embryonic development. This image shows the human embryo at two, three, four, five, six, seven, eight, twelve and fifteen weeks of development. From Ernst Haeckel, *Anthropogenie oder Entwickel ungs-geschichte des Menschen,* Leipzig: W. Engelmann, 1877, fig. 122.

Being asexual does not mean being celibate. Although female parthenogenetic lizards do not need sperm to conceive, many species still show vestiges of sexual behaviour—often mating with males of closely related species. For example, in the desert grasslands whiptail lizard *(Cnemidophorus uniparens)*, copulation may stimulate females to lay their eggs. Photo courtesy David P. Crews.

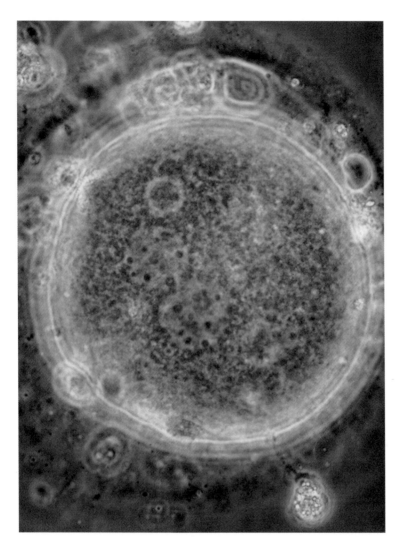

How we all start. A human egg and sperm. Scientists have long wondered why the two cells that create us are so very different. Photo courtesy The Carnegie Collection, National Museum of Health and Medicine.

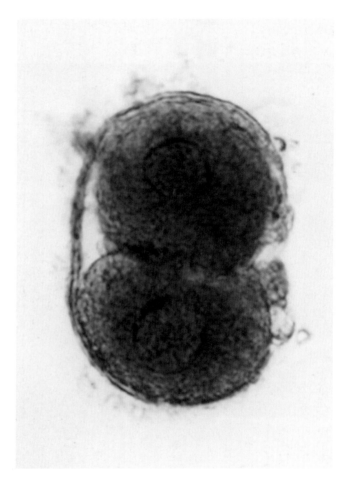

Within a day of fertilisation, the embryo rushes headlong into a frenetic round of cell division, forming first a two-cell embryo. Photo courtesy The Carnegie Collection, National Museum of Health and Medicine.

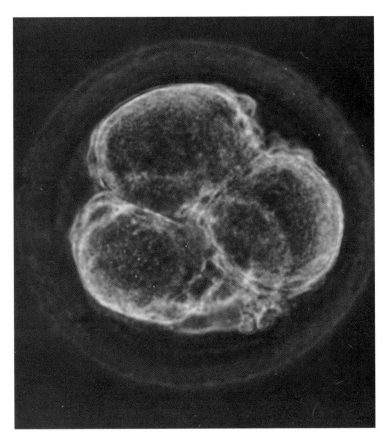

A human four-cell embryo (the fourth cell is hiding behind the other three). Photo courtesy The Carnegie Collection, National Museum of Health and Medicine.

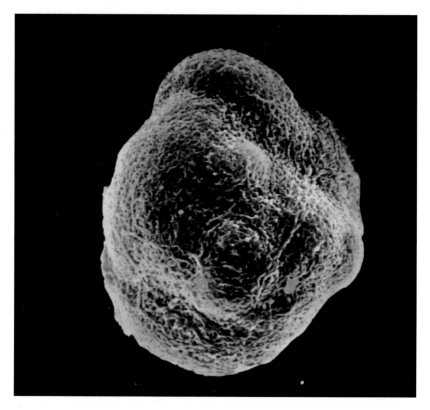

Within days of fertilisation, the human embryo has become a ball of cells, or 'morula'. At this stage, there appears to be little difference between the cells. Photo courtesy SPL/Custom Medical Stock Photo.

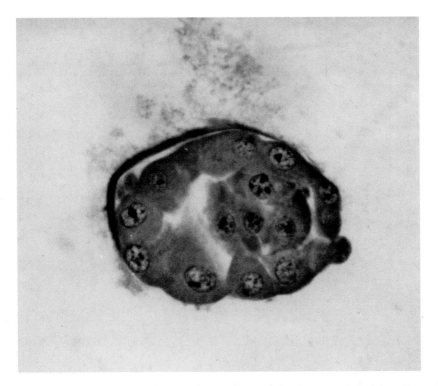

On the fourth day the embryo makes its first real developmental decision. By the blastocyst stage, the cells in the embryo have divided into two distinct groups: the outer trophoblast (a hollow ball of cells that will make the outer layer of the placenta) and the inner cell mass, part of which will form the baby. Photo courtesy The Carnegie Collection, National Museum of Health and Medicine.

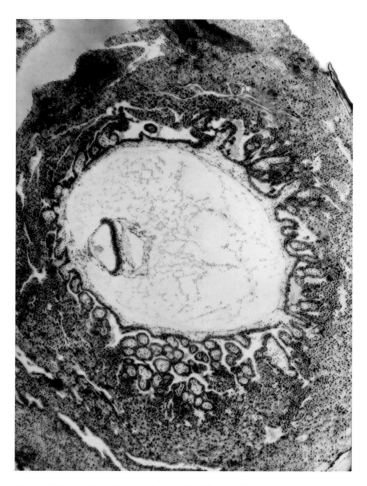

One of the main roles of the trophoblast cells is to invade the
wall of the uterus. In this image, the embryo can be seen floating
in a fluid-filled sac. This sac is lined by trophoblast cells, which
are invading aggressively into the surrounding maternal tissue.
Photo courtesy The Carnegie Collection, National Museum of Health
and Medicine.

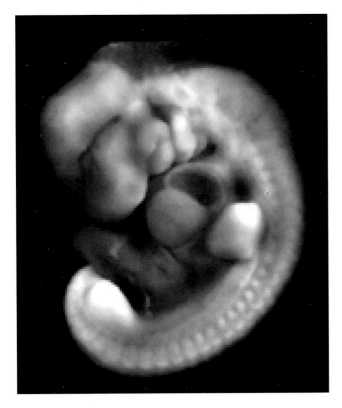

Four weeks after fertilisation, the human embryo is slowly
taking on baby-like form. At this stage it is dominated by
the large developing brain (top left) and the long segmented back (right, curling around to lower left). The paddle-
like buds of the arms and legs can be seen growing from
the back. Photo courtesy The Carnegie Collection, National
Museum of Health and Medicine.

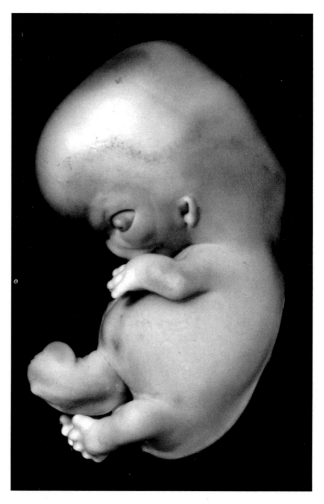

Side view of a human embryo after 7 weeks. Photo courtesy The Carnegie Collection, National Museum of Health and Medicine.

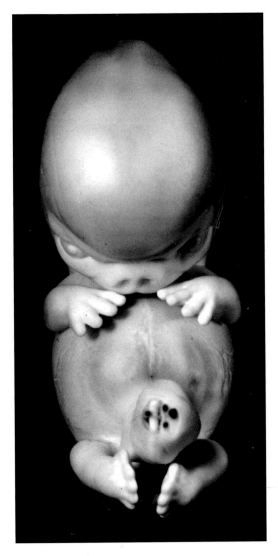

Front view of a human embryo after 7 weeks. Even by this early stage, the basic body plan of the baby has been established, and the facial features, fingers and toes are clearly visible. Photo courtesy The Carnegie Collection, National Museum of Health and Medicine.

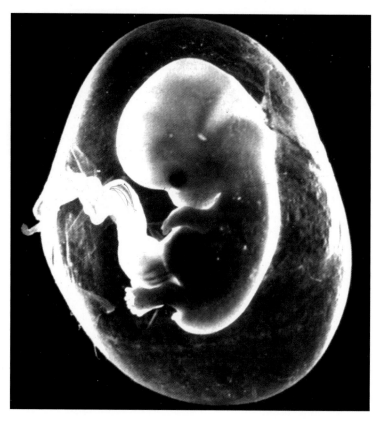

By eight weeks of development, the baby has now formed almost all
its internal and external structures, and by convention becomes
known as a fetus. Here the fetus floats within its inner, amniotic
membrane. The large pale umbilical cord can be seen passing from its
abdomen to the membrane. Photo courtesy SPL/Custom Medical Stock
Photo.

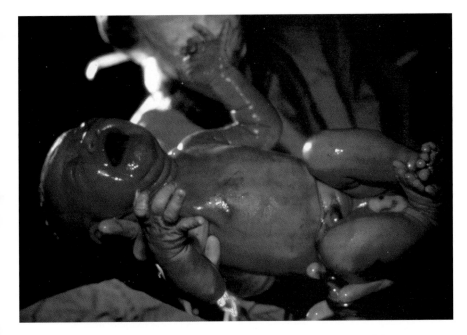

Birth is the greatest upheaval that we ever experience, and one of the most dangerous times of our life. As soon as a baby is launched into the outside world, it must adapt rapidly to its new environment. The challenges that face it include breathing air, disconnecting the placenta, keeping warm and starting to drink milk. Photo courtesy SJU Biomed Com/Custom Medical Stock Photo.

The development of the brain is relatively slow compared to the rest of the body. Over the first year of life outside the womb, babies gradually develop new ways to perceive, interpret and interact with the outside world. By the end of this year the child will have developed a remarkably mature personality and even a sense of humour. Photo courtesy David Bainbridge.

The sex chromosomes present us with a paradox. Although the role of the X and Y chromosomes is to produce two discrete populations of male and female humans, the way they are inherited means that babies can be born which are 'intersex': neither fully male nor female. A child must inherit most of its chromosomes in perfect pairs if it is to be healthy, but the rules about sex chromosomes are much more lax. There are few limitations on how many Y chromosomes a person can have – perhaps because an embryo's survival is not affected by whether or not it has a Y chromosome. Also, because any cell that has more than one X chromosome can just switch off any 'excess' X chromosomes, embryos can survive with one, two, three or even four X chromosomes.

So because of the body's relaxed attitude to the number of sex chromosomes a cell can have, babies with abnormal sex chromosome complements are usually viable, and some are superficially similar to 'normal' XX and XY people. These abnormalities usually result from abnormal loss of sex chromosomes or the acquisition of extra sex chromosomes, usually during the production of sperm and eggs or around the time of fertilisation. Whereas having too many or too few copies of other chromosomes is usually disastrous, a considerable number of the healthy human population have abnormal sex chromosomes. It has proved very difficult to get a good idea of exactly how common these abnormalities are, but intersex people may constitute 0.5 per cent of the adult population. Because these people are usually infertile, more cases are coming to light now that many couples are seeking help from fertility clinics.

Women with Turner's syndrome have only one sex chromosome, and so are said to be XO. They have female external genitals, but do not have normal ovaries and are completely infertile. They are usually short and have underdeveloped genitals and breasts as well as being prone to deafness and infertility. XO women often also have webbed skin between their neck and shoulders. Diagnosed cases of Turner's syndrome are probably the tip of a large prenatal iceberg. As many as 1 per cent of human embryos may be XO, but most of these die before birth, so that only around 0.03 per cent of the adult population has Turner's syndrome. There is no male equivalent of the syndrome: YO embryos do not survive because they lack all the essential genes present on the X chromosome.

By contrast, the 0.1 per cent of the population who are XXX are often indistinguishable from XX women, although they are more often mentally subnormal. This chromosome abnormality is probably the one that most often goes unnoticed, as XXX women are often fertile. This is even true of women with more X chromosomes – some with four or even five. The effects of having extra X chromosomes are probably subtle because these women simply inactivate any excess X chromosomes in the same way that they would normally inactivate a single X in each cell.

By this argument, however, the 0.2 per cent of the human population who are XXY men should appear normal too, but in fact such men often develop enlarged breasts and do not produce any sperm – a condition known as Klinefelter's syndrome. They also tend to have longer legs than XY men. As well as having supernumerary X chromosomes, men can also have extra Y chromosomes. The effects of men carrying XYY chromosomes are intermediate in severity: extra-Y men are often taller and may have learning difficulties. There have been reports that extra-Y men are more likely to become criminals, but other studies suggest that this is not the case. Klinefelter's syndrome and the extra-Y condition are just the two commonest forms of male intersexuality, and some men have been diagnosed as XXXY, XXXXY, XXYY or XXXYY.

All these examples of chromosomal intersex agree with our expectations about the functions of the X and Y chromosomes. Yet rare cases arise that do not seem to tie in with our ideas of how humans decide their sex. The first of these is the hermaphrodite, someone with both ovarian and testicular tissue, and these are usually XX (or more rarely XX/XY mixtures). Hermaphrodites tend to have their testicle on the right and their ovary on the left, although we do not know why. There is a second paradoxical condition called sex reversal, in which apparently normal men turn out to be XX.

Although these two examples do not agree with our simple model of sex determination, they have helped to establish that the part of the Y chromosome which confers 'maleness' is a discrete, small region. In hermaphrodites and sex reversal cases, it is thought that this region of the Y chromosome has been accidentally transplanted on to an X chromosome, now called an X^y. The 'mixed' nature of hermaphrodites presumably results from the random X inactivation of the abnormal X^y throughout the body, but we still do not know the mechanism underlying sex reversal.

The actual gene that makes embryos become male has recently been identified on the Y chromosome. Called *SRY*, this gene is the first single gene known to trigger the formation of an entire human organ system – the male genitalia. Of course, now we know the maleness-trigger gene, the race is on to find out how it controls the developing reproductive organs in such a profound way. What we already know is that the primary function of *SRY* is to form a testicle rather than an ovary.

Once the testicle forms, just about everything else follows. The testicles use two hormones to control the rest of the body. First, Müllerian Inhibiting Substance causes the Müllerian uterine ducts to regress – there would have been no need for a uterus had our baby been a boy. Second, androgens (male sex steroid hormones) start to carry out many of the more 'positive' aspects of maleness, such as the construction of the sperm-producing tubules of the testicle. Androgens also drive the conversion of the Wolffian ducts into the tubes that will store sperm and carry them to the penis.

Androgens control the formation of the external genitalia as well. Male sex steroids remould the slit-like aperture that drains the bladder to the outside world. First of all, the slit seals up and the exit tube is redirected along a projection at the front of the slit, the penis. As the penis grows, the tube grows with it and eventually opens at its tip. Meanwhile, an apparatus is established in the scrotal swellings to draw the testicles down into the scrotum from their position by the kidneys, although this may not take place until after birth.

Unfortunately, things can still go wrong at this stage, even if the

embryo has inherited a normal XX or XY chromosome complement. Sometimes the ovaries or testicles simply fail to form for
some unknown reason, and a set of underdeveloped female tubes
are left in their stead. Occasionally, XY males have testicles that
form all the usual pipework, but which then themselves regress.
Unlike these cases of gonadal intersex, sometimes the problem is
very clearly hormonal. The commonest causes of this are abnormal
generation of androgens in XX embryos, and failures of androgen
or Müllerian Inhibiting Substance activity in XY embryos. Males
that lack Müllerian Inhibiting Substance, for example, appear
normal except that they retain the Müllerian tubes.

By an exceptional genetic fluke, hormonal intersex is particularly common in one part of the world. In the Dominican
Republic, intersexuality is so common that many parents accept
that they can have three different types of baby: boys, girls and
'guevedoces'. Guevedoces look superficially like girls at birth, with
a small clitoris and apparently normal labia. The only sign that they
are any different from other little girls is that they do not have a
normal vulva – instead they urinate through their clitoris. This
state of affairs continues all through childhood, until things
suddenly start to change at puberty. Just when other girls are
starting to turn into women, guevedoces turn into men. Their
clitoris has a sudden growth spurt as it quickly transforms into a
normal penis, and their labia swell and sag until two normal
testicles drop into them. Rather poetically, the islanders call the
testicles 'eggs', so 'eggs at twelve': 'guevedoces'.

We now know what causes guevedoces. The condition is
inherited and it can happen anywhere in the world, but this one
island simply seems to have accumulated more than its fair share of
the genes that cause it. Guevedoces are XY and they have normal
testicles. These testicles make the androgen testosterone, and this
makes the Wolffian ducts develop into normal male internal
genitalia. However, guevedoces are unable to convert testosterone
into another androgen called dihydrotestosterone. This is the
hormone that makes the external genitals of fetuses become male,
and so guevedoces are born looking externally like girls. When
puberty comes, the testicles start to make much more androgen,

and this allows the penis and scrotum to finish their development, producing a 'normal' XY male.

One striking feature of guevedoces is that they often do not seem to find this natural sex-change process very traumatic. You would have thought that puberty was stressful enough, without having to change from a girl into a boy at the same time. Yet it seems that the social acceptance of this 'third type of baby' makes life much easier for guevedoces. Local doctors are adept at telling which little girls will be women and which will be men, and so there is usually little surprise when testicles and penis swing into view at puberty. Also, it is thought that guevedoces may be aware that they are different from most girls even before puberty. After puberty, guevedoces are almost indistinguishable from other men, and they often marry and have children – which is, of course, one reason why the condition remains common. In some ways, guevedoces are better off than other men – their unusual androgens mean that they do not get acne as teenagers, and as they get older they do not suffer from the usual curses of male middle age, pattern baldness and a misbehaving prostate.

The 'third sex' of the Dominican Republic raises important issues for dealing with intersexuality in other societies. In fact, the abundance of intersexes on that island has made it the one place where 'managing' intersexuality has not become a big issue. Elsewhere, the very rarity of intersexuality has often made life difficult, even intolerable, for intersexes. Until quite recently, intersexuality was treated as a disorder that required treatment, preferably as a matter of urgency. It was believed that children that did not fit neatly into 'male' or 'female' categories should be surgically modified so that they looked like one of these two mainstream sexes. There was a widespread assumption that early assignment to one sex or another was infinitely preferable to suffering the social stigma of childhood micropenis or clitoromegaly. As a result, parents were often never told the exact cause of their child's intersexuality, and were instructed instead to reinforce the new surgically created sexuality of the baby, and even to misinform their children about their original sexuality.

The problem with this approach is that simply telling a child that it is a certain sex does not seem to be enough to make it 'feel' that it is that sex. Babies were modified into boys or girls mainly because making them look like one sex was surgically easier than making them look like the other, and many later found that they did not 'look' like they 'felt'. Some believed themselves to be male, but had no male genitals, and some felt like women, but had a penis. Also, many are unable to enjoy any form of sex because of the surgical disruption of their genitalia – they complain of being mutilated. Many psychiatrists now recommend a much more 'hands-off' approach to intersexual children. They now believe that surgery should be carried out only where a child's anatomy makes it run the risk of getting other problems, such as infections. The child should be raised as neither girl nor boy until it is able to express its feelings about its own sexuality – after puberty, some intersexes want to make themselves more male or more female, but many do not. Many do not feel the need to be forced into being male or female at all. Male, female; normal, abnormal; treatment, mutilation: intersexuality makes all these concepts seem much more fluid.

Little of this is relevant to our little embryo, however, because she is an XX girl, with all the usual hormones. Around day 45 she starts to form the cells that will create the ovarian follicle and cumulus cells around the egg. Her second-kidney fades away, taking its ducts with it. Instead, estrogen from the ovary starts to stimulate the Müllerian uterine ducts to grow and join together, starting at around day 50. The uterus is formed by the fused parts of the ducts, while the unfused parts make the Fallopian tubes. Ovarian hormones also drive the development of the female external genitalia. The slit that would have closed up in a boy remains open to form the vulva, and the bump that became the penis grows somewhat less exuberantly and becomes the clitoris. Finally, the swellings that in boys receive the testicles and become the scrotum, retain their original conformation in girls and form the labia majora.

Meanwhile, the ovary is a hive of activity. In a remarkable example of biological preparedness, girls make all their eggs before they are born. The female sex cells lose their amoeba-like behaviour once they reach the ovary and start to proliferate to produce thousands of primordial eggs. Once made, these eggs then settle down to wait for two or three decades before they are reawakened for ovulation.

There is tremendous wastage in the ovary, however, as most of the thousands of eggs made by the female fetus never even get as far as ovulation. Most of the eggs that a baby girl makes are dismantled before she is even born, and a continuing process of attrition eats away at her reservoir of eggs right up until puberty. Unlike men, who can churn out sperm willy-nilly – if you will excuse the phrase – women are limited to using only the eggs they are born with. This, of course, raises the possibility that eggs can deteriorate during the long wait. However, one thing is certain: this all confirms what a unique thing the human egg is – not only can it generate a whole new person in a way that no other cell can, but it is also prepared to wait for twenty to forty years for the opportunity to do so.

When things go wrong

The first seven weeks of an embryo's existence are extremely hectic. The embryo changes more during this short time than she will change in the whole of the rest of her life. A great deal has to be done, and the embryo's future is in doubt if a single step in her great construction project goes wrong. If her genetic instructions are faulty, or the conditions for her development are not perfect, now is the time that problems usually become apparent. After day 50, life will get easier and the embryo will spend most of its time just growing. Life for the early embryo, however, is lived on a knife edge between successful development and death.

This is why failing pregnancies usually fail within the first eight weeks. Although it is difficult to be precise, it has been suggested that around 75 per cent of human embryos die before the eighth week of pregnancy. Another reason why pregnancies may fail at the beginning rather than at the end is that this is the least disruptive way to deal with embryonic death. Presumably it is far better to discard a faulty embryo before its mother has invested eight months of her time and energy in it. Obviously, early pregnancy loss is also less psychologically devastating for human parents than late miscarriage, especially if an embryo is lost even before a woman knows she is pregnant. Yet this still does not explain why so many human embryos die. By comparison, in Chapter 2 we saw that most red deer hinds can reliably produce a healthy calf from a single egg every year. When the high rate of pregnancy loss in humans is combined with women's inability to know when they are fertile, you can see that human reproduction is hardly designed for maximum output.

Although embryonic death is extremely difficult to study in humans, we now think that embryos tend to die during the most demanding stages of their existence. Perhaps 25 per cent of human embryos die in the first week, even before they have implanted. Embryos have a lot to do at this time, and many simply do not seem able to cope. They must mix their parents' genes correctly, they must start cell division, they have to start using their own genes and they must then prepare for implantation. The second week does not seem much easier, as embryos concentrate on implanting into the wall of the uterus and stopping their mothers' cycles. Around another 25 per cent may fail at this stage. Perhaps another 25 per cent fail over the next six weeks – as they suddenly find that there is some terrible reason why they cannot meet their construction deadlines.

The commonest cause of embryonic death during the first third of pregnancy – the first trimester – is thought to be abnormal chromosomes. Although the number of sex chromosomes is not critical, embryos must inherit exactly forty-four other chromosomes, or things go wrong. Most embryos with extra or missing chromosomes die during the first trimester, although there are exceptions. Down's syndrome embryos acquire an extra chromosome, but the extra chromosome they have is unusual in that it is not lethal, so Down's babies survive to term and often into adulthood.

Chromosomal abnormalities are also a reason why older mothers tend to lose more embryos. Because older women's eggs are themselves older, they seem to be more likely to make mistakes in how many chromosomes they try to contribute to the embryo. Because of this, chromosomal abnormalities are commoner in the offspring of older women, and so early pregnancy loss is also more common. There are probably also other reasons why older women lose more embryos, but these are not entirely clear – perhaps the ability of the uterus to welcome embryos deteriorates with age, or maybe older women are more prone to diseases that adversely affect pregnancy.

Another cause of embryonic death is multiple pregnancy. Even the presence of a twin can reduce an embryo's chances of survival, but the risks become greater as the number of competing embryos sharing the uterus increases. In recent years, this problem has presented obstetricians with a dilemma. Because so many embryos die after fertility procedures such as *in vitro* fertilisation, it is common practice for doctors to transfer many embryos into a woman's uterus to maximise the chances that just one will survive. Unfortunately, this procedure can work too well and multiple pregnancies of between five and eight embryos have recently become far more common. The sheer number of siblings in these pregnancies usually places the entire pregnancy at risk, and doctors usually recommend a procedure called selective fetal reduction, in which several embryos are removed before the pregnancy fails. Of course, selective fetal reduction is a very controversial procedure, especially as these high-number multiple pregnancies do not always fail. Although the statistics are clear that fetal reduction does, in general, improve the chances of some babies being born, this cannot be guaranteed for any individual pregnancy.

So the first fifty days are a time when the embryo risks everything in her headlong rush to develop. Certainly, by the end of the first trimester, most of the danger has passed. Her chances of surviving the second trimester are good. By the time she reaches twenty-six weeks, advances in neonatal medicine now mean that she would probably survive if she were born. It is almost as if, by surviving her first seven weeks, she has proved that she can succeed.

Instead of the ball of cells we left at the end of the last chapter, we now have a baby girl. As her priorities change from arranging her organs to simply growing, she changes from an embryo to a fetus. However, she has not been alone while she has been developing. Far from it – she has been in constant contact with her mother. The baby is in a delicate position. She is absolutely dependent on her mother to support her and not to reject her like some unwelcome invader. All the clever developmental planning in the world would come to nothing if she lost her mother's cooperation. In the next chapter, we will see how the baby has constantly to mollify her distrustful landlady to achieve this cooperation. This tiny parasite must now persuade its mother to accept it as her own.

○ ○ ○ ○ ○ ○ ○ ○ ○

*Why does the mother
accept the 'foreign' baby?*

○ ○ ○ ○ ○ ○ ○ ○ ○

CHAPTER 4

The visitor within

∘ ∘ ∘ ∘ ∘ ∘ ∘ ∘ ∘

Mammals have been doing something subversive for 90 million years. At great risk to ourselves, we have contravened one of the great laws of life: never let outsiders in.

All through their history, animals have tried to stop themselves being colonised by other organisms. Sometimes they fail, and this is why many creatures today make a good living as parasites. Our life has become one long struggle against invading freeloaders, and animals have developed immune systems to defend themselves. The complexity of the immune system is testament to animals' obsession with keeping the outside world out: honed over millions of years, every day, every creature's immune system keeps a constant vigil for approaching invaders.

This sensible strategy of excluding foreigners from our bodies had to be turned on its head when we became mammals. Suddenly, our females wanted to increase the protection they could afford their offspring by becoming long-term baby incubators. Female mammals now had to learn to break the golden rule of the immune system and let a foreign being in, instead of throwing it unceremoniously out. Babies are not the only foreign organisms that make it through the defences into a woman's body, but they are the only ones that are actively encouraged in.

The precarious nature of this mammalian system became apparent only when doctors tried to carry out organ transplant surgery. Once again humans expected their own bodies to accept a foreign interloper, just as women do when they become pregnant. As a result, we now know that people's bodies simply do not like having bits of other people spliced in. But why doesn't a woman's body react in the same way when an embryo tries to transplant itself into her uterus? Why is the fetus not rejected like a tissue graft?

The story of the mammalian 'fetal transplant' is one of the strangest parts of our journey through pregnancy – the mammalian fetus is an exception to the rule by an exceptional group of animals. Why is it accepted by its mother? When we study pregnancy, we are looking not at conflict, but potential conflict. There is no wholesale maternal–fetal war.

The new frontier

Just like a real war, the pregnancy cold war centres on a front line where the two protagonists uneasily face each other across no man's land. The frontier between our baby girl and her mother is, of course, the placenta. The developing fetus depends on its mother for its food, oxygen and sewerage, so it cannot just seal itself off from her. There must be contact for the fetus to keep itself alive, so the placental interface is where our baby now has to persuade its mother to accept her.

In the previous chapter we saw how the inner cell mass moulds itself into a baby, but we largely ignored the hollow bubble-like trophoblast that encloses it. The trophoblast is the first specialised tissue ever produced by the human embryo, and it is destined to form the outer membrane that encloses the baby, and the outer layer of the placenta. The trophoblast will be the part of the fetus that will be directly exposed to its mother. Now we will see how very important the trophoblast is: as the baby's outer shroud, it is her physical link with the outside world.

Unappetisingly, the word 'placenta' is Greek for 'pudding', and the ancient Greeks were among the first to realise its importance. The first European to suggest that the role of the placenta was to feed the developing baby was Diogenes of Apollonia, who claimed that it is present as the baby changes from a formless mass into a recognisable infant. Diogenes suggested that babies feed by sucking from their placenta and that this is why they produce meconium (fetal feces) soon after birth. He even claimed that a baby's months spent sucking the placenta explain why it is so adept at suckling at its mother's breast as soon as it is born.

The trophoblast springs into action as soon as the embryo comes to rest on the wall of the uterus. The trophoblast cells immediately embark on a spectacular bout of multiplication and soon split into two populations. The population on the inside of the trophoblastic bubble remains as individual cells, the cellular trophoblast, but the outer cells lose their partitioning cell walls and join together to make the fused trophoblast, which is in direct contact with the mother. The fused trophoblast is also the tissue that produces the all-important hCG to arrest the mother's menstrual cycles.

The fused trophoblast uses an unusual mechanism to join up into a non-cellular bag. Recent experiments have provided convincing evidence that trophoblast cells are forced to fuse by proteins very similar to those produced by viruses. For some time, viruses have been known to produce proteins that help them to fuse with cells so they can infect them. Yet for the last decade, scientists have also wondered why genes for these viral proteins can also be found scattered among our own genes. We now believe that, at some time in the past, these genes were 'left behind' by viruses that inserted some of their genes into human cells. Once incorporated into our ancestors' cells, these abandoned genes were passed on to subsequent generations, and eventually to us. Every human carries several of these discarded viral genes, and until recently no one knew if they did anything. One of these genes has now been found to make fusion proteins in the fused trophoblast, so it appears that human babies now use one of these abandoned viral genes to form their placenta. So the human baby steals a trick from a virus so that it can itself become a parasite.

The fused trophoblast is extremely invasive, and it chews aggressively into the uterine wall until the embryo is completely buried. Its savagery is a particular feature of humans – most other mammals' trophoblasts are much better-behaved guests. The infiltration of human fetal tissue into the mother's uterus has even been compared to the invasion of normal tissue by malignant cancers.

The invasiveness of human trophoblast leads to problems when it tries to invade in the wrong place – a condition called ectopic pregnancy. Around 1 per cent of all pregnancies implant in the wrong place, and ectopic pregnancy is becoming more common. The commonest site for a misguided embryo to implant is in the Fallopian tubes, and embryos probably come to rest here because they are prevented from migrating down to the uterus after fertilisation. Infection is a common cause of blockages of the tubes, as are failed attempts to use the 'morning-after pill'. The morning-after pill is a potent mix of steroid hormones designed to make the Fallopian tubes and uterus contract to expel the embryo, but sometimes the morning-after pill makes the Fallopian tubes stop moving altogether and the embryo has to make the best of a bad job and implants where it is. The embryo's sheer drive to survive allows it to attach, invade and form a placenta wherever it is – ectopic pregnancies can also attach to the cervix, ovary or another abdominal organ. Ectopic pregnancies cause pain and erratic bleeding and can be fatal if left untreated: as the embryo grows, it can rupture the Fallopian tube, leading to massive loss of blood.

If it is fortunate enough to lodge itself in the wall of the uterus, the trophoblast bubble will start to grow, along with the embryo developing inside it. The growing embryo-bubble causes the uterine wall to bulge ever further out into the uterine cavity until it abuts the opposite wall. In the course of time, the entire uterus swells to fill the pelvis and then it takes over most of the abdomen as well. As it does so, it displaces the bladder, intestines, stomach and liver, and eventually even squeezes the heart and lungs in its relentless take-over of the mother's body. Many of the mother's discomforts of pregnancy result from this gradual edging-out of her organs by the ever-growing uterus: incontinence, colitis, low blood pressure, indigestion and breathlessness are added to the joys of morning sickness, varicose veins and sheer bulk.

The side of the trophoblast bubble that first invaded the uterus now becomes the placenta. The placenta is a specialised thickening of the fused and cellular trophoblast membranes, as well as the thickened wall of the uterus. The job of the placenta is to allow exchange of nutrients and waste between mother and baby. Simple contact of fetal and maternal membranes is insufficient to meet the baby's needs, however, and the placenta soon makes several alterations to allow it to function more efficiently. The fetal membranes first increase their surface area dramatically by developing thousands of finger-like projections that penetrate deep into the wall of the uterus – the more area for exchange of goodies, the better. The baby then establishes a profuse blood supply to these trophoblast fingers, which will carry nutrients from her placenta through her umbilical cord into her body. Her mother reciprocates by constructing an extensive blood supply system on her side of the placenta. To keep the placenta full of blood, the mother adds some extra water to her blood to increase its volume, as well as to dilute it so that it can flow more easily through the blood spaces in the placenta.

Human mothers make a further, especially altruistic, concession to their babies. Because the placenta works by passing chemicals between the maternal and fetal blood, the thickness of tissue through which these chemicals pass is important. Human mothers sacrifice large portions of their side of the placenta to allow their babies easy access to their blood. At the start of pregnancy there are six layers of tissue separating the maternal and fetal blood in all mammals: the cells lining the mother's blood vessels, some fibrous cells that support the uterine lining, the uterine lining itself, the fetal trophoblast, fibrous cells supporting the trophoblast and cells lining the fetus' blood vessels.

As we have seen, the fused trophoblast erodes through the uterine lining extremely rapidly. This invasion continues until the maternal fibrous tissue is obliterated as well, so that nothing remains between the fetus and the maternal blood vessels. At the same time, within the fetal fingers, the baby's own blood vessels are edging her own fibrous tissue out of the way to give them direct access to the trophoblast membrane. Finally, as the cellular trophoblast takes over the role of invader from the fused trophoblast, something remarkable happens: the baby's trophoblast cells crawl through the lining of her mother's blood vessels and come into direct contact with her blood. These fetal cells then replace the mother's blood vessel cells until most of the maternal blood spaces are actually lined by fetal cells. This replacement of the mother's blood vessel walls causes them to collapse into flaccid bags filled with the mother's slow-flowing blood. The trophoblast fingers now waft about, luxuriating in a pool of vivifying maternal blood. By a kind of trophoblast *Blitzkrieg*, the fetal placental cells have wiped out all of the maternal side of the placenta, leaving just two of the original six layers of tissue separating the mother's blood from the baby's – the trophoblast and the fetal blood vessels.

This obliteration of the maternal side of the placenta is a characteristic of humans, but we are not sure why the baby needs to do it. This is an important issue for the baby's survival because her annexation of the placenta has left her own tissue dangerously exposed to her mother's xenophobic gaze. How does the baby benefit from this invasion if it places her in such jeopardy?

Biologists have tried to gain insights into the invasive behaviour of human embryos by looking at other animals' placentas. As with so many other aspects of reproductive biology, there seem to be almost as many ways of making a placenta as there are species of mammal. For a start, many mammals do not have a single, circular placenta like ours: some monkeys have two circular placentas, calves have more than a hundred mini-placentas scattered all over the uterus, and kittens and puppies are wrapped in a circular placental belt. The variation does not stop there. The way the fetus attaches to the uterine wall varies between different mammalian species as well. Many species form placental fingers like we do, but others throw the trophoblast membrane into folds, branching folds or even an interlocking labyrinth of tubules intermeshed with the maternal tissue.

Yet studying all these other species has made one thing clear: there is no obvious reason why the human placenta has to be so invasive. While our close primate relatives and some rodents share our placental invasiveness, and a few species of bat have even more aggressive trophoblasts than ours, other species cope perfectly well with three, four, five or even a full six layers of tissue separating the maternal and fetal bloodstreams. Take the horse, for example. When a foal attaches to a mare's uterus, little tufts of trophoblast interlock neatly with the uterine wall. There is no erosion of the mother's tissue over most of the placenta and all of the original layers of the placenta remain intact. Despite the six layers separating foals from their life-support machine, they survive perfectly happily for the full duration of a leisurely eleven-month pregnancy.

The inexplicably aggressive behaviour of the human trophoblast causes considerable problems for the uterus, which has had to be redesigned to cope with the ravages of its invader. Unlike the inside of a mare's uterus, which is a simple membrane with a few glands, the human uterus is built up into a thick buffer zone to absorb the ferocity of the baby's attack. This baby-buffer zone is the material shed every month at menstruation. It also gets torn away with the placenta at birth, and because of its disposable, replaceable nature, the buffer zone is called the decidua. The deciduate nature of the human placenta contrasts with many other animals' placentas, in which the baby's side of the placenta separates neatly from the mother's at birth. This is why there is little blood spilt during calving, foaling, lambing, kindling or whelping. For some reason, women are unlucky enough to have to deliver babies determined to peel a layer off the inside of their womb when they are born.

The placenta: sustenance and sewerage

Now the baby is fully 'plumbed in' with her brand new placenta, she needs all the things that you and I need: food, oxygen and a way of getting rid of some waste products. Conveniently enough, because the baby uses the same nutrients as the mother's own tissues, the mother's blood is constantly laden with everything the baby needs. All the baby has to do is select what she needs and leave the mother to replenish the larder. Similarly, the waste products made by the baby are identical to those that the mother's own body generates – the baby needs only to connect to the maternal mains sewerage system and all her excretions will be automatically washed away. The advantages of this system are twofold. First, the mother does not have to make any special arrangements to supply exotic foodstuffs to her baby. Second, the baby gets used to running a postnatal-like metabolism while still in the womb – birth will be stressful enough for the baby without her having to change her metabolism completely.

The baby uses different methods to obtain the different things she wants. The placenta is well designed, and the method used to transport a particular nutrient usually reflects how quickly it can drift across on its own, how much of it the baby needs and whether she can cope with having less of it in her blood than her mother.

Small nutrients, such as salts, can usually get across the placenta by simple diffusion. For example, if there is more sodium in the mother's blood than in the baby's, the tiny sodium atoms can drift across the placenta to make up the shortfall on the fetal side. This process actually works so quickly that the levels of salts on each side of the placenta are usually very similar. There are other types of food molecule that cross the placenta in this way, such as fats and fatty vitamins (A, D and E). These are not as tiny as salts, but they can easily permeate the fatty membranes that line the cells of the placental barrier.

Glucose seems to be a special case and it has its own special way to cross from mother to baby. The fetus needs a lot of glucose as it is her main energy source, but glucose does not cross cells very quickly. The baby's solution to this problem is called facilitated diffusion. Built into the membranes of the placental cells are special 'dumb-waiter' molecules that bind glucose on one side and move it across to the other. These molecules can actually pass glucose in either direction, but because it is more abundant on the mother's side, they tend to collect glucose there and pass it to the baby. The great advantage of this dumb-waiter system is that it does not need any energy to work – all the baby has to do is to make the dumb-waiter molecules and they do the rest.

The baby has to go to greater lengths to get hold of some other constituents of her diet. If a nutrient is especially reluctant to cross the placenta (such as amino acids or the water-soluble vitamins B and C), or needs particularly careful handling (like calcium, iron or iodine), then the placenta actively pumps that molecule from mother to baby. Like dumb-waiter molecules, pumping molecules are set in the placental membrane, but unlike the dumb waiters they need energy, because they usually pump nutrients 'uphill' from the low concentration of the mother's blood to the high concentration of the baby's blood.

One thing that the baby needs in large quantities is, of course, oxygen. To get all the oxygen she needs, the baby has developed perhaps the cleverest transport system of all, designed to extract the maximum amount of oxygen from her mother's blood without expending any energy at all. Oxygen diffuses across the placenta quite readily, but the baby is also able to fiddle its oxygen accounts by judicious use of haemoglobin. Unlike most of the baby's requirements, oxygen does not float free in the bloodstream, but is bound to the red blood pigment haemoglobin. Haemoglobin is a protein that absorbs molecules of oxygen where it is abundant and releases it where it is scarce. The fetus cannot breathe through her lungs and so instead she absorbs oxygen through her placental 'lung'.

The baby's first trick is to cram more haemoglobin into each drop of blood than her mother does. This means that her blood can carry more oxygen to her tissues for each of her little heartbeats. Placental ingenuity goes further than that, however, because the baby's haemoglobin-rich blood is also able to absorb more oxygen from her mother's blood at the placenta as well.

Babies increase their oxygen gains by using a special type of haemoglobin. Fetal haemoglobin is very like the haemoglobin pumping around you and me, but it does have one subtle, but very important feature: it absorbs oxygen slightly better. This increased affection for oxygen is not important when it comes to delivering oxygen to the baby's tissue, nor will it make much difference when the baby gasps its first breaths. The place where improved oxygen binding comes into its own is the placenta, where it helps the baby to wrest oxygen from its mother. At the placenta, fetal and maternal haemoglobin are almost in competition for oxygen and the greater desire of fetal haemoglobin for oxygen allows it to draw it away from maternal haemoglobin. Fetal haemoglobin pleads poverty to its mother's haemoglobin, almost feigning the appearance of an oxygen-starved tissue. By the simple innovation of a different haemoglobin molecule, the baby makes her mother relinquish more and more oxygen.

The fetus' ingenuity knows no bounds, however, and one more sleight-of-hand maximises her ability to extract oxygen from her mother. This trick relies on the carbon dioxide she is continually producing as her main waste product. Carbon dioxide is excreted into the mother's blood through the placenta, but it serves one more useful function before it leaves the scene. Carbon dioxide is very slightly acidic when dissolved in water or blood, and so when the baby offloads it at the placenta, her own blood becomes more alkaline while her mother's blood becomes more acidic. Haemoglobin makes use of this by being sensitive to the acid/alkaline balance of the liquid surrounding it. When the blood gets more acidic, haemoglobin tends to shed oxygen; and when blood gets more alkaline, haemoglobin likes to absorb more oxygen. Just as the fetal blood is shedding carbon dioxide, the oxygen-binding capacity of its haemoglobin is increased, and the opposite happens on the maternal side of the placenta. So, with one last grab, the baby wrings a few more molecules of oxygen out of her long-suffering mother.

The placenta pulls off a real *tour de force* of biological economy, efficacy and elegance. Beautiful it is not, but this couple of pounds of flesh is one of nature's most underestimated inventions. Unfortunately, disposability is the feature of the placenta that sticks in most people's minds. Almost as soon as it appears, it seems, the placenta is put in the bin. This shabby neglect of our ugly 'other halves' is largely a western conceit, and in more reflective cultures the placenta has been elevated to mystical heights. When a child is born in Bali, for example, its placenta or *ari-ari* is washed in perfumed water, wrapped in cloth and interred by the threshold of the family home in a ceremonially prepared coconut. In fact, the *ari-ari* is considered one of the spiritual siblings of the child, and respectful treatment encourages it to start its life-long job of protecting the infant from misfortune. Perhaps this is a rather more fitting end for a placenta than to be incinerated in a little yellow plastic bag.

The baby meets its mother's immune system

Whether or not the Balinese are right and an *ari-ari* can protect a baby after it is born, it certainly protects it before it is born. As I said at the start of this book, pregnancy is a uniquely intimate association between two people, and now we can see that most of this intimacy takes place at the placenta. The problem with all this unusual consorting between two different individuals is that, while it allows mothers to support their babies through a prolonged pregnancy, it also leaves the baby vulnerable to attack by its mother. Usually, women are as eager as anyone to launch attacks against foreign invaders, but there is a degree of restraint in this aggressive self-defence when a woman becomes pregnant. There is no evidence that women ever reject their babies like a tissue graft. In fact, the horse is the only natural example of a mammalian female that launches a rejection reaction against part of its offspring's placenta, and even that reaction never seems to compromise the success of equine pregnancy. So why do women not reject their babies?

You may think I am overstating the potential of the fetus to provoke immune attack by its mother, but the immune system is specifically designed to hunt down and destroy foreign materials throughout the body. As any transplant recipient will tell you, the immune system will go to remarkable lengths to find and kill alien tissue. Most researchers believe that the fundamental function of the immune system is to distinguish 'self' from 'non-self' and to destroy the latter. All through the animal kingdom, creatures have developed methods to reject foreign material, presumably because it may be a potential parasite.

The ability to maintain the 'integrity of self' seems to have evolved almost as soon as cells first clumped together into many-celled animals. The drive to recognise and exclude non-self cells is the constant theme of immunity throughout the animal kingdom, and the only difference between different animals is the ways in which they choose to achieve this exclusion. In jellyfish and corals, for example, the immune system can remember foreign materials encountered in the past, so it can react to them more rapidly if it meets them again. As we will see, this immune 'memory' is a particularly important characteristic of our own immune system. Flatworms and molluscs have dedicated immune cells that patrol the body looking for interlopers – again, specialised immune cells are a central part of our own defences. Insects have developed two further weapons also found in the human armoury. The first is cells that engulf and destroy foreign material and the second is the ability to seal up particularly recalcitrant invaders into harmless packages.

Really, we have learnt about the immune system of non-mammals as an afterthought. The history of immunology has not been a neat progression starting from the study of simple animals and progressing to complex ones, or even a concerted attack on the question of how the body distinguishes self from non-self. Instead, our knowledge has accumulated in a rather jumbled fashion – an observation of a disease here, a description of a patient with immune deficiency there. This unfocused approach explains the course of the next part of our story: why much of twentieth-century immunology was directed at understanding the capricious results of transplanting tissue from one person to another. It did not become clear until much later that the uniquely human activity of transplanting organs might tell us a great deal about how animals detect and destroy foreign invaders – the mechanisms we use to reject transplants are the same as those that attack invading micro-organisms.

Even then, there was yet another twist to the story. Before scientists had fully understood how transplantation and infection are linked, some had already realised that the mammalian fetus might be an exception to the rule that animals recognise and destroy invading foreign organisms. Whatever theory was devised to explain transplant rejection and immunity to infection, it must also account for the fetus' apparently privileged status.

Hard graft: the discovery of the laws of transplantation

As long as people have attempted surgery, they have dreamt of being able to replace damaged organs by transplantation. During Christ's lifetime there were claims that St Peter had used his God-given powers to replace the breasts of a young woman mutilated by the Romans. Miraculous grafting continued in the third century when St Cosmas and St Damian allegedly transplanted a leg from a 'dead moor' to replace a diseased leg. By the sixteenth century, doctors were attempting non-miraculous transplantation – a Dutch writer reported that bone from a dog's skull had been used to repair a damaged human skull. No doubt there were many other ancient and medieval attempts at organ transplantation that produced results so disastrous that the embarrassed perpetrator never recorded them. Even as late as the nineteenth century, French surgeons were attempting to connect the heads of guillotined prisoners to the bodies of decapitated dogs.

Our modern understanding of the mechanisms of transplant rejection was one of the few benefits of the global wars of the twentieth century. Just as technology was allowing martial carnage to reach new levels of efficiency, medical science was finally giving surgeons the opportunity to treat, rather than amputate, the casualties of war. Medical technique was now clean enough to allow transfusion of blood from one individual to another. Also, surgeons confronted with severe skin burns had started experimenting with transplanting skin between individuals. It soon became clear, however, that transfusion and transplantation were not as straightforward as their exponents had hoped. All too often, and apparently at random, things went wrong.

Doctors quickly realised that blood transfusions either succeeded or failed miserably – there did not seem to be much middle ground. Some recipients would see their blood loss quickly remedied, but for others the outlook was grim. Transfusion from 'inappropriate' donors led to an uncontrollable and often fatal reaction in the recipient, as if the foreign blood was being 'rejected'. What doctors needed was a method for accurately predicting the outcome of blood transfusions. Experience showed that some individuals were fortunate enough to be able to accept blood from any donor, whereas others could receive transfusions from only a subset of the human population.

Although the mechanism was mysterious, rules started to emerge about who could receive blood from whom. In fact, the ability of people to receive blood successfully from each other depends largely on whether just two proteins are present on their red blood cells. When early experiments on blood typing were carried out, no one knew what these proteins actually were, so they simply called them A and B. We still use these letters to describe blood groups today. If someone has both proteins, then they are said to be blood type AB. If they have just one, then they are classed as either A or B. If they have neither, then they are type O – there is no 'O protein', it just means you have neither A nor B proteins.

Your ability to accept blood from a donor depends on whether you and the donor share any A or B proteins. If you are transfused with blood carrying the A protein, for example, and you have no A, then you will recognise the A as 'foreign' and reject the transplanted blood cells, with catastrophic results. This means that if you are AB, then you are lucky, because no blood you can be given will carry a protein that you have not 'seen' before. Alternatively, if the donor is O, then their blood carries no A or B protein to which your immune system can react (this is why doctors in television medical dramas are always calling for 'O-negative' blood).

This simple system governing blood transfusion taught us three things about transferring tissue from one individual to another. The first is that sometimes tissue can be transferred successfully. The second is that acceptance of tissue depends on the presence or absence of a very small number of cell molecules – in this case two, A and B. The third lesson is that we inherit these transplantation-governing molecules from our parents in a simple Mendelian fashion.

When surgeons tried transplanting tissues other than blood, however, they found that the A and B system was no help at all – other tissues seemed to behave differently. Transplanting these tissues was a much greater practical challenge not only because the mechanism that controlled rejection was mysterious, but also because tissue rejection could work both ways. Transplant recipients can reject their new organs (host-versus-graft rejection), but the graft can also try to reject the recipient as well (graft-versus-host rejection). The problems of early transplant surgery seemed almost insuperable. The road to today's routine and highly successful transplant surgery was long and arduous, but it gave us compelling insights into how we distinguish self from non-self.

Before the Second World War, little was known about the machinery underlying transplant rejection in humans. It was strongly suspected that the factors controlling rejection were genetic because it was known that 'identical' twins could usually accept tissue grafts from each other – they did not appear to distinguish between self and almost-identical non-self. Apart from this nugget of information, the whole problem seemed much more obscure than that of blood transfusion. Non-blood transplants did not seem to work at all.

A great step forward was made by the Brazilian-born Anglo-Lebanese zoologist Sir Peter Medawar, professor at the universities of Oxford, Birmingham and finally London. When Medawar tried to carry out repeated transplantation procedures between animals, he noticed that an intriguing pattern emerged. First, he confirmed what surgeons already knew: that tissue transplanted from a donor animal to an unrelated recipient is usually rejected. He also showed that, if tissue is subsequently transferred to the same recipient from a different donor, then another, similar rejection reaction occurs. Crucially, however, he also demonstrated that if a third transplant is carried out from the first donor to the same long-suffering recipient, then the rejection reaction is faster and more intense than either of the other reactions.

It seemed to Medawar that the recipient's body had 'remembered' that a graft had been carried out previously from the first donor and reacted with renewed vigour at the second attempt. The recipient had not been made more sensitive to transplantation in general, but just to transplantation from the specific donors that had contributed tissue before.

This apparent in-built 'recall' of which donors had been used in the past was a feature of transplant rejection that fascinated Medawar, partly because biologists had seen animals 'remember' exposure to foreign materials once before. For some decades, scientists had known that animals' immune systems respond in just such a way to injected bacterial poisons: animals react more strongly the second time that a particular poison is administered, regardless of what other bacterial material is injected in the interim. To Medawar this was clear evidence that rejection of transplants is carried out by the immune system – the same machinery that protects the body from invading microbes. This realisation, in the early 1940s, was to win him a Nobel Prize in 1960.

Although we take it for granted today, this sudden joining of transplantation and the immune system was a crucial point in medical science. It told us what immunity actually is. The common ground shared by rejection of microbes and rejection of grafts is central to all our modern ideas of how bodies protect themselves, but like all major advances it raised as many questions as it answered. Scientists still could not say if the body reacts to foreign grafts in exactly the same way that it reacts to bacteria and viruses, nor did they know if there is some unifying feature of grafts and microbes that makes the body treat them in the same way. There was one other issue that perplexed early transplantation biologists. There seemed to be very good reasons why animals should have developed methods of attacking foreign microbes, but no one could explain why they are so good at attacking foreign organ transplants. There was no obvious reason why animals should have evolved a way to reject transplants, because animals never encounter transplants in nature.

Despite all these questions, there remained a more immediate problem: whether doctors could now use the new link between graft rejection and the immune system to allow them to carry out rejection-free transplant surgery. The immediate solution to this challenge was not intellectually pleasing, but it was, literally, a lifesaver. There was such a great need for transplantation that doctors did not have time to recreate the perfect matching of donor and recipient that had solved the problem of blood transfusion. Unlike the A and B proteins of blood, no one knew which molecules dictated the success of non-blood transplants. Instead, early attempts at transplantation focused on suppressing recipients' immune systems in the post-operative period, to stop them attacking the transplanted organ. This 'damping down' of the rejection process with immunosuppressive drugs is far from an ideal solution, but in the absence of perfectly matching donors, surgeons have little choice. Even today, when we know which molecules cause transplantation reactions, immunosuppression remains an important part of transplantation medicine.

The discovery that the immune system is responsible for graft rejection makes our baby's predicament more clear. If we think of the baby as a potential graft, then we now know that she must appease her mother's immune system if she is to survive inside her. To understand how she does this, we will have to find out what it is about tissue grafts that makes the immune system react to them so strongly. As we will now see, tissue grafts are marked out as 'foreign' by just a few special molecules that they carry on their surface. Presumably, the baby will have to be extremely careful with these master transplantation molecules if she is to have a secure future.

The MHC: controller of graft rejection

After the identification of the link between transplantation and immunity, the next step was to identify the molecules that cause transplantation rejection. Their discovery in the 1950s has completely changed how we think about our body's defences, and they have become a central thread in our theories of how babies avoid rejection by their mother.

Like the A and B molecules that control rejection of blood transfusions, the molecules that control transplantation rejection are few in number and are inherited in a simple Mendelian fashion. Yet they are far more important than A and B because they are present on almost every cell in the body (in fact, red blood cells are among the few cell types that do not carry the master-molecules and this is why rejection of blood is unusual). The central role of these molecules in graft rejection has earned them the name 'major histocompatibility complex' or MHC ('histo-' means 'tissue').

The MHC molecules are really very special, and almost everything we have discovered about them has turned out to be unexpected. In fact, studying the MHC has become a scientific discipline in its own right. Human cells can carry up to six pairs of different MHC molecules, with one of each pair made from a gene inherited from one parent. For example, MHC-A, MHC-B and MHC-C are present on almost every cell in the human body, and MHC-DP, MHC-DQ and MHC-DR are found on specialised immune cells. The presence of MHC molecules on almost every cell in the body is one reason why almost any transplanted organ can induce a rejection reaction.

The MHC genes, and the proteins they make, have some unusual features of great importance to our baby. These features are all related to the role of the MHC in protecting the body, and they even help the MHC act as the final arbiter of self and non-self. Graft rejection, avoiding infection, having a baby – these things are all distilled into the MHC molecules.

Soon after they were discovered, researchers started to study whether the MHC genes have any effect on the diseases that people get. They discovered that a person's MHC genes have a powerful effect on their chances of suffering from a wide range of different diseases, from diabetes to rheumatoid arthritis to malaria. We hear a great deal about scientists trying to find the genes linked to individual diseases, but the MHC genes are implicated in an enormous array of different conditions.

Another unusual feature of the MHC genes and the proteins they produce is that they are extremely variable within the human population. For example, human MHC-A has more than a hundred different forms. This degree of variation is very unusual – most genes are present in just one, two or three forms. Because MHC genes are so variable, when a recipient receives a graft from a donor, it is an odds–on bet that it will carry MHC proteins that the recipient's immune system has not seen before. This is why organ transplants are usually rejected as 'foreign'. Unfortunately for the baby whose pregnancy we have been following, she is in the same position as that tissue graft. If she produces MHC proteins from genes that she inherited from her father, then her mother might see them as 'foreign', and reject her.

Animals seem to take special care of how their MHC genes are inherited. In one study in which researchers allowed a colony of mice to breed as they wished, female mice seemed to prefer to mate with males with different MHC genes to their own. Female mice achieve this trick because they can tell what MHC genes a male mouse is carrying by the smell of his urine. The effect of this female preference for MHC-dissimilar males is that most baby mice get different MHC genes from each parent. It also helps to maintain the large number of different MHC genes in the population.

Of course, once someone had discovered that mice choose their sexual partners on the basis of their MHC, it was only a matter of time before someone tried to find out if the same was true of people. The nearest human equivalent that the Swiss scientists who carried out this study could find to a freely breeding colony of mice was a population of university students. Rather than actually observing who copulated with whom, they instead carried out an odour preference test. Male students were asked to wear a T-shirt for two nights, and women students were then asked to score the odours for attractiveness. Remarkably, in a bizarre echo of what female mice do, the women tended to prefer T-shirts that had been worn by men who carried different MHC genes from themselves. Also, the odours that reminded women of their previous sexual partners also tended to be those of MHC-dissimilar men.

This raises the possibility that women may choose their mates partly on the basis of their MHC genes. It is almost certain that, like mice, men's odours do depend on their MHC genes, because rats can be trained to discriminate between people with different MHC genes on the basis of their smell. This does not, of course, mean that women can do the same thing. Yet perhaps we are underestimating our sense of smell – just because we find it very difficult to describe or remember smells does not mean that we are not very good at detecting them. Maybe women really do choose their partners partly because of their exotic MHC-associated odour.

The Swiss study also discovered that women taking the contraceptive pill show completely different preferences – if anything, they prefer the smell of men with the similar MHC genes to themselves. The researchers suggested that the pill creates a hormonal state similar to pregnancy. Indeed, female mice appear to change their preferences for companions when they become pregnant – they too then prefer MHC-similar males. All this raises the question of how the contraceptive pill affects women's mate choices. If women are designed to choose mates that ensure that their children get a good mix of MHC genes, what will happen to the MHC genes in the human population now that many women select their partners when 'under the influence' of oral contraceptives?

A few studies have been published which may explain why women are keen to avoid having children with men with similar MHC genes. Different research groups have claimed that couples with similar MHC genes may be more prone to repeated mis-carriage. Also, *in vitro* fertilisation attempts may be more likely to fail in MHC-similar couples. The link between partners' MHC genes and pregnancy failure is not universally accepted, but if it is a real effect, it may go some way to explaining women's choice of partners. We do not know why an embryo with identical copies of its MHC genes should be at such a disadvantage, but this does not mean that it is not. The MHC seems to affect so many aspects of human life, there is no reason to assume that it cannot control pregnancy too.

Mother's immune defences

So: placentas are exposed to it, transplants are rejected by it and the MHC is pivotal within it. Indeed, the immune system is a triumph of biological engineering, a triumph on which we each depend for our survival. All day, every day, each of us is assaulted by a barrage of tiny organisms trying to turn us into their home, and the immune system is what keeps them out. The immune system is one of the most complex components of the entire body, and I will only scratch the surface of that complexity.

When a microbe first tries to gain entry to the body, it comes up against the simplest, but perhaps the most effective, part of the immune system. The internal and external surfaces of the body are sealed by a covering that resists most attempts at invasion by alien creatures. This covering can take the form of waxy skin, stomach acid, antiseptic tears or the little hairs in your windpipe that waft debris out of your lungs, and all of these barriers conspire to make the human body a difficult place to enter.

For the few organisms that do penetrate the body's outer lining, life becomes even more difficult once they are inside. The body is constantly scoured by legions of 'guard cells' (called granulocytes and macrophages) for evidence of an intruder. These guard cells are alerted to danger by a few simple clues – they are activated by chemicals present on the surface of many different bacteria, or by the debris that spews forth from damaged body tissues. Evolution has primed them to react when they detect these tell-tale signs of infection and to hunt down and destroy the organisms responsible.

A very small number of microbes are able to enter the body and escape the attentions of the guard cells. There is one more line of defence for these hardy invaders, and this is the lymphocytes. Formed in the bone marrow, lymphocytes are just about the cleverest little cells in the whole body, and they are our only defence against many terrible diseases. Remarkably, they can detect almost any foreign material, regardless of whence it came. Almost all of the myriad alien creatures that enter the body can be recognised by at least one lymphocyte, and when they are recognised, a horrendous assault is unleashed on them. The few lymphocytes that have detected the foreign invader are driven into a frenzied round of proliferation that soon leads to a veritable army of cells, each dedicated to killing the invader.

One set, the antibody lymphocytes, makes special antibody proteins that latch on to the invading organism and stop it working properly or trigger other cells to destroy it. Another group of lymphocytes, the thymic killer lymphocytes, have molecules on their surface that lock on to cells infected by the micro-organism so that they can kill them. (They are called 'thymic' because, after leaving the bone marrow, they spend time developing in the thymus, the organ that we mentioned developing from the gill grooves in the previous chapter.) A final group, the thymic helper lymphocytes, coordinate the attack and also lie dormant after the infection is over so that an even more ferocious attack can be mounted should the microbe ever enter the body again. This in-built recall of the organisms that have attacked the body in the past is the feature of the immune system that Medawar noticed was similar to the way that animals can remember which donors they have received skin grafts from.

So the lymphocytes are formidable opponents, each potentially able to generate an entire army of cells that can overwhelm most invaders. Yet their amazing abilities have led immunologists to wonder how they can respond to all of the almost infinite number of possible organisms to which they can be exposed. To make it easier for lymphocytes to respond to foreign material in infected cells, these cells prepare this material before they present it to them. They do not present whole organisms – instead they chop up the organism's proteins into short fragments and present these to lymphocytes. There are far fewer possible types of short fragment than there are potential types of foreign organism, so this chopping-up makes it much easier for lymphocytes to decide whether they are being presented with 'self' or 'non-self' material.

The whole of transplantation biology fell into place in the early 1970s when it was discovered that these foreign fragments are served up to the immune system on MHC molecules – this is the actual biological function of MHC proteins. Lymphocytes are therefore attuned to binding MHC molecules carrying protein fragments. This is why the MHC is so important: it is the final common pathway by which the tremendous variety of alien molecules provoke immune responses. Lymphocytes do not 'see' molecules in isolation. Instead, they 'see' them displayed against the backdrop of an MHC molecule.

We can now see why transplant rejection is carried out by the immune system, why it is dependent on the MHC and why it is so aggressive. Imagine what happens immediately after a tissue graft. When a cell within the transplanted organ tries to present fragments to the recipient's immune system, the recipient's lymphocytes will immediately realise that the MHC on the graft is foreign. So a very large number of the transplant recipient's lymphocytes will react against the MHC on the transplant and this

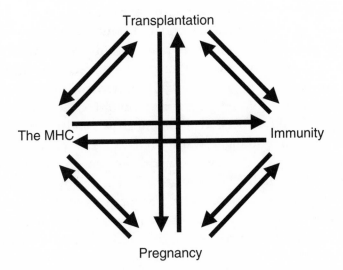

Figure 5. The immune predicament of the fetus. The immune system is responsible for rejecting transplanted organs, and it does this because it reacts to MHC (major histocompatibility complex) molecules on these transplants. We now realise that, to avoid attack by its mother, the fetus must pacify her immune system by carefully concealing MHC molecules from her. By doing this, it reduces its chances of being rejected like a transplanted organ.

is why transplant rejection is so aggressive. Far more lymphocytes react to foreign MHC than to any foreign microbe.

With the discovery of the function of the MHC, scientists could finally explain how and why the immune system rejects both micro-organisms and transplanted organs. They were also now in a position to work out how the developing fetus avoids being rejected. Suddenly, transplantation, immunity and the MHC were all mixed together with pregnancy (Figure 5).

So how does the baby hide herself?

We can now bring together all we have learnt to tell us why our baby is not rejected by her mother. There are many tempting pieces of information which look like they might fit into an all-explaining story, and perhaps that story will soon be complete.

In the first half of the twentieth century, some immunologists and reproductive biologists wondered how the mammalian fetus might have cheated the maternal immune system. How do mammals manage to retain their young internally, bristling as they do with 'foreign' paternally inherited molecules? The question was first placed on a firm footing by Medawar in 1953. He realised, like others, that mammalian mothers were doing something rather strange. What Medawar did differently, however, was to suggest three ways in which babies and mothers might coexist successfully. First, he suggested that a baby might remove provocative molecules from the placenta to stop it being recognised as foreign. Second, he proposed that there may just be some sort of physical barrier that keeps the mother's immune cells away from the baby. Finally, he raised the possibility that the fetus may suppress its mother's immune system in some way. As valid today as they were in 1953, these three suggestions still form the central canon of pregnancy immunologists' thinking.

When pregnancy immunology first got going in earnest, there was not much evidence that women are immunosuppressed during pregnancy. Women did not seem to be taking this risk just to protect their babies, so the maternal immune suppression idea was put on the back burner. Similarly, in humans, in which fetal placental cells are directly exposed to the maternal immune system, there did not seem to be a physical barrier between the mother and fetus. So the physical barrier theory was also edged to one side. Instead, most scientists chose the other option: that the baby fiddles with the molecular make-up of its placental cells to avoid detection. But were they right?

Switching off Dad

Genetic imprinting is, as we have seen, a genetic trick that affects how a small number of special genes are used. Instead of using both copies of these genes, imprinting means that only the copy inherited from one particular parent is used. So imprinting might offer the fetus a way to avoid exposing all those problematical paternal molecules to her mother.

Genetic imprinting would be a clever solution to the fetus-as-graft problem because it would not compromise the mother's ability to fight infection. Simply switching off all the paternally inherited MHC genes on the placenta would mean that there would be no need for the mother's immune system to be suppressed. It would hardly be to the fetus' advantage to render her host a playground for all sorts of bacteria and viruses. Instead, without paternal MHC, she would look just like part of her mother's body. The mother's lymphocytes would then tolerate the child's presence in the same way that they tolerate any part of her own body. And she would be at no more risk of infection than if she were not pregnant.

One piece of evidence that supports the idea that babies avoid immune attack by switching off paternal genes in the placenta is the finding that most genetic imprinting takes place before birth. Conveniently enough, imprinting is active at exactly the time when the baby has to evade its mother's immune system. Imprinting appears to have little effect during adult life, and normal imprinting is, as far as we know, primarily a prenatal phenomenon. Even better, imprinted genes are often used in the placenta and fetal membranes.

This is all very well, but does it prove anything? Just because imprinting is a good idea and happens to be taking place at roughly the right time does not mean that it plays any part in protecting the baby. Indeed, there is evidence accumulating that imprinting of genes in the placenta may have a completely different purpose. It is now thought that imprinting is actually part of a struggle between mother and father to control fetal growth. When a woman conceives a child, she must balance two opposing aims: the baby should grow as rapidly as possible, but she must not become such a drain that the mother's reproductive future is compromised. By contrast, fathers want big healthy babies and do not care so much for their mate's future ability to produce children – those future babies might not be theirs anyway. It has been proposed that a struggle has developed between mothers and fathers respectively to limit and enhance the growth of the fetus in the womb. This struggle is fought by paternal growth-promoting genes that are switched on in the fetus, and by the mother switching off the copies of these genes that she contributes to the fetus. The idea that imprinting is involved in the control of fetal growth is supported by the fact that maternal genes are more often switched off in the placenta than paternal ones. This makes sense if these genes encourage fetal growth, but it does not fit in with the idea that fetuses 'imprint away' paternal MHC genes. For this, you would expect that mostly paternal genes would be switched off.

In the late 1990s, scientists tried to resolve the question of genetic imprinting by looking to see if the MHC genes are imprinted in the placenta. Contrary to expectations, paternal MHC molecules are, in fact, very much in evidence on the surface of the placenta, flagrantly dangling into the mother's bloodstream. The lack of imprinting of the MHC genes almost certainly means that imprinting is not used to protect the baby from her mother's immune system. So we continue our quest by looking in more detail at the transplantation molecules that the baby allows on to the placenta.

Protect and survive

When scientists first looked for MHC on the human placenta in the 1980s, they had a surprise. Unlike almost every other tissue in the body, the placenta does not have any of the usual MHC molecules on its surface. In lacking these otherwise ubiquitous molecules, the placenta is a very unusual tissue indeed. The implications were obvious: the fetus may avoid graft-like rejection simply by switching off the proteins that might make it look like a tissue graft.

Although the lack of all the usual MHC molecules on the placenta seems a good way to avoid maternal immune rejection, a problem springs to mind – this system does not seem very safe. MHC proteins are made in almost all other cells in the body, and they are made for a reason – they allow infected tissues to raise the alarm by presenting fragments of microbes to lymphocytes. So, by discarding its MHC molecules, the placenta could be putting itself at risk of uncontrollable infection.

At this point in the history of placental immunology, the human baby played a wild card. In the late 1980s a new MHC molecule was discovered on the placenta, called MHC-G. Although slightly smaller, it seemed similar to the more usual types of MHC. The placenta has apparently switched off its more humdrum MHC just so it can switch on an 'exotic' one instead. The most striking feature of MHC-G was, and still is, the fact that it is found in just one place in the human body: the placenta (or, more specifically, the outer layer of the fetal membranes and the fetal cells that burrow deep into the maternal tissues). Tantalisingly, the distribution of MHC-G is almost the complete opposite of the distribution of the 'usual' MHCs – they are present on almost all tissues except the placenta. It is hardly surprising, then, that the discovery of MHC-G was greeted with much speculation about whether it is the key to fetal survival.

Almost as soon as MHC-G was discovered, the hunt was on to find out why the molecule has been posted at the immunologically uneasy frontier between mother and baby. An important piece of the MHC jigsaw fell into place in the 1990s when geneticists looked at MHC-G in the human population. It soon became clear that the MHC-G gene does not show the tremendous variation characteristic of other MHC genes. Whatever has caused MHC-G to lose its variability, the implications of this loss are clear. If the variation in the other MHCs is what ensures incompatibility between transplant donors and recipients, then the lack of variation in MHC-G could allow the 'fetal graft' to be accepted. Presumably, if MHC-G is invariant, then mothers will have MHC-G molecules identical to those on the placenta, so they will not reject their babies as they would reject any other 'transplant'.

In the late 1990s, it became clear that MHC-G may also solve the baby's other problem – how to maintain vigilance for microbes when all the 'usual' MHC genes are switched off. MHC-G can carry protein fragments just like other MHC proteins do when they want to present foreign material to lymphocytes. Also, MHC-G may be able to interact with lymphocytes like other MHC proteins do.

MHC-G may also protect the baby in another way. One type of

immune cell dominates the lacklustre ranks on the maternal side of the placenta – a kind of guard cell called the 'natural killer'. This discovery has led to the idea that MHC-G's main job may be to fend off natural killer cells. Like all immune guard cells, natural killers have preconceived ideas about what to attack, and natural killers make a habit of attacking cells with very little MHC, usually tumours. Trophoblast is, therefore, a potential target for natural killers as well, but MHC-G is able to stop natural killer cells attacking it. So MHC-G has another role – convincing the maternal immune system that the placenta is not a tumour.

But why should a baby act like a transplant?

The story of MHC-G seems to go a long way towards explaining how the fetus keeps itself safe from maternal immune attack. Recently, however, many researchers have questioned the approach of treating pregnancy as a suppressed transplantation rejection, and this new approach has some justification. For example, how can babies be considered as potentially 'rejectable' tissue grafts when they are apparently never subject to immune attack? In fact, in many ways the fetus behaves rather unlike a tissue graft.

The argument about whether the fetus can be thought of as a graft may simply reflect the fact that mammals solved the immunological problem of pregnancy a long time ago, so the fetus is no longer at risk of rejection. Obviously, the ancestors of modern mammals cannot have been prone to rejecting their unborn offspring, or we simply would not be here. Yet this does not mean that there was no immunological problem for early mammals to overcome as pregnancy evolved. Presumably, it was the gradual improvement of mammals' adaptations to pregnancy that rendered the modern human fetus so safe from its mother's immune system. Unfortunately, the efficiency with which the baby is protected is now the greatest obstacle to researchers trying to find out how that protection is achieved. The mammalian mother's acceptance of the child within her is now so complete that it is hard to see how it ever evolved.

Another problem with studying the fetus as a transplant is that we do not fully understand what happens when an ordinary graft is rejected. Although transplantation has been the subject of decades of research, there are still great gaps in our understanding of what drives graft rejection. It is unreasonable to expect to be able to understand the subtleties of pregnancy if we cannot explain the consequences of crude artificial transplantation. Just as we do not know why pregnancy is such an immunologically benign process, we also cannot explain why transplantation rejection is so ferocious.

A fetal armada invades the mother's blood

There is another feature of human pregnancy that makes the placenta seem even less like an ordinary organ transplant. It is now known that the placenta is not the sole frontier at which mother and baby have to resolve their immunological differences.

Every human mother is constantly exposed to a second source of fetal provocation – cells that leak from the baby into its mother's circulation. In recent years, more and more evidence has accumulated that the placenta is far from a perfect barrier between mother and child. The mother is continually assaulted by a stream of fetal blood cells and placental fragments dribbling into her bloodstream. Many of these cells are probably not as circumspect as placental trophoblast cells and it is likely that many of them have a full complement of MHC proteins.

This raises an important question: if these cells are available for maternal immune inspection, what is the point of the placenta's careful control of its MHC? It is difficult to explain why the fetus should go to all the trouble of carefully switching off MHC molecules on the placenta if a continual dribble of MHC-bearing cells is entering her mother's blood. Perhaps we have been looking at the placenta in the wrong way – maybe it is not designed to avoid the maternal immune system seeing it at all. Recognition of the fetus may be unavoidable because of the stream of random fetal cells percolating indiscreetly into the maternal bloodstream. It seems that there is no way that the fetus can stop the mother detecting paternal MHC – instead the placenta's job may be to withstand the resulting immune attack.

Indeed, there is good evidence that mothers' immune systems do detect their developing fetuses, and they probably do this because of the cells that fall off or leak through the placenta. Many pregnant women's lymphocytes make antibodies that recognise their babies' paternally inherited MHC, but these antibodies do not seem to affect pregnancy in any way. The failure of these antibodies to damage the fetus may have something to do with the structure of the placenta itself. Possibly the placenta acts as a barrier to antibodies that bind MHC proteins because it does not itself bear any orthodox MHC molecules to which they can attach. Alternatively, any antibodies that do get through the outer layers of the placenta may get 'mopped up' in the MHC-bearing tissue underneath. So the placenta may be acting like a version of Medawar's physical barrier between mother and baby.

The fetal cells that leak indiscreetly across the placenta may become very useful in the next few years. Modern techniques are sensitive enough to detect these cells as they drift around in the maternal bloodstream, and it is even possible to examine their genes. It has, for example, proved possible to determine the sex of human babies from fetal cells collected from maternal blood. This raises the possibility that we may soon be able to diagnose genetic diseases from blood samples from pregnant women. Older methods of genetic diagnosis needed samples collected from inside the pregnant uterus, but obstetricians now realise that this is not a risk-free procedure. So we may have acquired the ability to detect fetal genes in maternal blood at the perfect time.

The baby takes control

Medawar's third suggestion, that babies may adjust their mothers' immune systems to protect themselves, has once again come to the fore. Women do not become completely immunosuppressed during pregnancy and the reasons for this are obvious, but it is now thought that maternal immune defences may instead subtly change during pregnancy.

The first evidence for this alteration came when doctors studied what happens to women with autoimmune diseases when they become pregnant. In autoimmune disease, the immune system misguidedly starts to recognise part of the body as foreign and attacks it. Just like 'normal' immune responses, autoimmune diseases can be caused either by antibody cells or by thymic killer cells, and each autoimmune disease usually involves the undesirable activation of one of these two cell types. Intriguingly, doctors noticed that women with autoimmune diseases caused by antibody cells often showed a worsening of their condition during pregnancy. Conversely, women with diseases caused by thymic killers often reported improvements in their symptoms. This led to the suggestion that the baby might be shifting its mother's immune response away from thymic killers towards antibodies. But why?

There are several possible reasons why a baby might want to realign its mother's immune system from killers to antibodies. It might be because transplantation rejection is usually caused by thymic killers. Alternatively, it could be because antibodies are less dangerous to the fetus because her placenta acts as a barrier to antibodies. Whatever the reason for the shift in the maternal immune system, it is clear that the baby is not acting like a 'simple' tissue graft: organ transplants do not control their recipient's immune system. The immunological shift can also be seen as yet another attack by the baby on her mother's biological independence. Not content with orchestrating a complete reorganisation of her mother's anatomy and physiology, the baby now wants to fiddle with her immune defences as well. This is not just an idle comment: it has been suggested that the killer-to-antibody shift may actually put mothers at greater risk of infection by microbes tackled by thymic killer cells – a nasty bunch including leprosy, malaria, tuberculosis and toxoplasmosis.

Birth is not the end

Although the baby does all she can to make her mother's immune system accept her, or at least ignore her, she can never be entirely free of the risk of maternal attack. Remarkably, she is still at the whim of her mother's immune system even after she is born.

It is ironic that human babies make all that effort to stop their mothers recognising their provocative MHC transplantation proteins as foreign, and yet the main immune threat to most babies is that their mother might react to a simple blood transfusion molecule. The 'rhesus' molecule is a protein present on red blood cells, just like the A and B proteins. Perhaps 90 per cent of people have the rhesus molecule, but it does not seem to be as important as A or B for successful blood transfusion. Pregnancy is the time when rhesus protein suddenly becomes dangerous, and if the mother reacts to her baby's rhesus molecules, that baby could get jaundice, anaemia or brain damage, or might even die.

Rhesus disease happens in women who have no rhesus protein themselves, but who react to rhesus on their baby's blood cells. Fetal blood cells often spill into the mother's blood, and this usually happens at birth when up to two tablespoonfuls of fetal blood can seep in. Less often, fetal blood can also leak through during pregnancy, and leakage is more likely during a miscarriage, or if the placenta forms in the wrong place. Because women usually react to rhesus protein at birth, however, their first rhesus–positive child is usually safely removed from its mother's immune system by the time it has mounted a full-blown response to rhesus protein.

The big problem with rhesus comes when a rhesus–negative woman is pregnant with her second, third or fourth rhesus–positive child. By that time, her immune system has been repeatedly exposed to rhesus protein and her antibody lymphocytes are pouring out large quantities of antibody, which can cross the placenta and stick on to her baby's blood cells. The baby's immune system thinks that anything coated with antibodies must be foreign, and so it starts to destroy its own blood cells. Not only may the baby lose so much blood that it becomes anaemic, but the breakdown of its blood haemoglobin creates large amounts of 'bilirubin' – a chemical that is toxic in large quantities.

During pregnancy, bilirubin is not too much of a challenge for the baby – it simply disappears across the placenta into the mother's blood. After birth, however, the baby has a problem. The maternal antibodies that crossed the placenta and stuck to its blood cells are still making it destroy its own blood, but it now has to get rid of the waste bilirubin itself. Newborn babies' livers are not especially good at getting rid of bilirubin, and it often accumulates for several days after birth, making many healthy babies become slightly jaundiced – the yellow colour of jaundice is due to bilirubin visible through the skin. Babies with rhesus disease are in a much worse position: they are producing far too much bilirubin for their liver to get rid of and it soon builds up to poisonous levels. Until recently, the bilirubin accumulation of rhesus disease was one of the most common causes of deafness, brain damage and death in newborn babies.

Only recently have we discovered how to prevent the disastrous effects of rhesus disease. We can treat the disease, by transfusing rhesus-negative blood into babies before or after birth, or by 'phototherapy' – sunlight and some artificial lamps alter bilirubin in the skin to a form that can more easily be removed by the baby's liver. More importantly, we know how to prevent rhesus disease starting in the first place. Women are now routinely tested for their rhesus blood group, and the blood group of their baby can be predicted from the father's group, or tested directly from the blood that drips from the baby's umbilical cord after birth. If a woman is considered to be at risk, she can be treated to reduce her chances of ever mounting an immune response to rhesus protein. At-risk women are injected with an antibody to rhesus protein, and this binds to any fetal blood that has leaked into them, so they destroy it – just like a rhesus disease baby does. The method works well because treated women destroy the fetal blood so quickly that they do not have time to make any antibodies of their own. So, if they are injected every time they have a baby, they are extremely unlikely ever to inflict rhesus disease on one of their children.

Although we know the mechanisms that cause rhesus disease, we do not really know why human pregnancy has evolved to allow it to happen. If 10 per cent of people are rhesus negative, then 9 per cent of couples can be expected to be a rhesus-negative woman and a rhesus-positive man. Why have we not evolved ways to protect these couples' babies? We simply do not know why rhesus causes the problems it does – after all, mothers often make antibodies that bind to their babies' MHC molecules or A and B molecules, but they do not seem to do any harm. The human fetus takes all its clever steps to stop itself being rejected by its mother, but it then allows itself to be fooled into attacking its own blood. A cynic might suggest that first-born babies are deliberately putting their younger siblings at risk by immunising their mothers to rhesus protein. Yet, to do this, they must carry a gene that will eventually put their own offspring at risk. Rhesus disease simply does not make sense.

Recently, evidence has accumulated which suggests that the maternal immune system may have yet more long-term effects on the fetus. These effects, if they exist, would be as difficult to explain as rhesus disease, as they seriously affect the chances of the fetus ever producing its own children. Homosexuality has long been a challenge to evolutionary biologists because they cannot explain why it occurs. Although many animals show homosexual behaviour, it is often simply misdirected heterosexual behaviour – male giraffes may court and even mount each other, but this is often just a prelude to fighting. In humans, however, homosexuality is a life-long sexual orientation, often to the exclusion of all heterosexual behaviour. Homosexuality dramatically reduces the chances of people passing on their genes to the next generation, and so it is difficult to explain why it is so common.

A number of different research groups have noticed a strange statistic about homosexual men: they tend to have more older brothers than heterosexual men. The results seem to suggest that every older brother makes a man 33 per cent more likely to be homosexual. This seems to be a very specific finding: homosexual men do not necessarily have more older sisters, and homosexual women are not thought to have any particular arrangement of siblings. Of all the theories that attempt to explain why older brothers have such a dramatic effect on male sexuality, the most intriguing involves the maternal immune system. If you accept that a man is already destined to be homosexual by the time he is born (and that is a big 'if'), then the older-brother statistic can only be explained by a biological system that 'remembers' the number of male babies the mother has already produced at the time she produces another son. And as we have seen, the immune system is one of the few parts of the body that 'remembers' things.

Male babies do indeed have extra proteins to which their mothers' immune systems can respond. One, called H-Y, is encoded by a gene on the Y chromosome and is especially prevalent in the developing brain. Pregnant animals are known to make antibodies to H-Y, presumably because this male molecule is entirely foreign to a female. This theory is based on the suggestion that, just as mothers who have had lots of rhesus–positive babies produce antibodies to rhesus protein, mothers who have had lots of sons make antibodies to H-Y. Just as in rhesus disease, these antibodies might cross the placenta, but, instead of sticking on to blood cells, they stick on to the regions of the brain that control sexual behaviour and preference. By binding to these areas, they may change for ever the sexuality of the fetus.

Before we rush headlong into accepting that homosexuality is the sexual equivalent of rhesus disease, I must emphasise that this is just a theory. Men do indeed have special molecules and there are good reasons to believe that human mothers can detect them as foreign and make antibodies to them. If such antibodies were made, there seems little reason why they could not cross the placenta (unless they got soaked up by H-Y molecules in the placenta, just like antibodies to MHC might). However, there is no evidence that they either stick to the brain or affect its development in any way. Yet the theory still has some appealing features. It could explain why homosexuality is such an apparently random occurrence, and why it is so common. Also, rather charmingly, it is a modern immunological version of the old idea that men become homosexuals because their mothers are too overbearing.

Reproductive immunology is still at the challenging stage when questions outnumber answers, and all we can say with certainty is that the fetus' solution to its immunological predicament is complex and subtle. Suffice it to say that the baby can and usually does coexist happily with its mother for nine months, in some ways intimately entwined and in other ways held very much at arm's length.

I want to add a single codicil to this chapter on the immunology of pregnancy. This is because it concerns a novel, medically important development that takes us back to some of our more mystical preconceptions about the placenta.

Diseases of the bone marrow are extremely important in modern medicine because they are common and they are debilitating – all the cells of the immune system are made in the bone marrow, so bone marrow disease leaves the body open to overwhelming infection. Marrow is fragile and it can be destroyed by a variety of diseases. Autoimmune diseases can strike at bone marrow just as they can strike anywhere else, and marrow is also prone to growth of cancers, which can obliterate all the normal marrow tissue.

For many of these diseases, bone marrow transplantation is the best treatment – wiping the slate clean and replacing old diseased marrow with new healthy cells. The key cells that need to be replaced are the progenitor cells that give rise to all the different types of blood and immune cell. Unfortunately, because marrow cells carry lots of MHC molecules on their surface and marrow can mount graft-versus-host reactions against transplant recipients, marrow transplantation is fraught with problems. Tissue matching between marrow donors and recipients must therefore be especially assiduous to reduce the chances of rejection. Half the battle for patients with bone marrow disease is won by finding a suitable matched donor, but new findings indicate that a perfectly matched donor may be found in an unexpected quarter.

In the western world, the umbilical cord and placenta are usually discarded after birth, but the discovery that the blood contained in the cord carries blood progenitor cells has generated new hope for transplantation medicine. There are not many of these cells, but it may be possible to grow them up to large numbers in the laboratory. Could they be collected at birth and preserved for use in later life? Currently, there is much controversy about the ethics of using fetal tissue to treat diseases of adults, but surely there is no reason why we should not use our own fetal blood, frozen since our birth. Perhaps a new ritual will soon be added to the western birth process – the technological ritual of rescuing and preserving our umbilical cord blood. Perhaps we will inter our *ari-ari*, not by our doorstep, but in deep freeze, as a benevolent protective force awaiting our cancerous hour of need.

○ ○ ○ ○ ○ ○ ○ ○ ○

*How do mother and baby
survive birth?*

○ ○ ○ ○ ○ ○ ○ ○ ○

CHAPTER 5

The visitor without

○ ○ ○ ○ ○ ○ ○ ○ ○

For most of human history, the process of birth has been fraught with risk for both mother and baby. Until recently, childbirth was one of the most dangerous episodes of a woman's life. Evolution had left her a baby with a huge head and expected her to expel it through a pelvis that had only recently been adapted for walking upright. Before the advent of modern midwifery and obstetrics, women probably had a 1 or 2 per cent chance of dying each time they gave birth. The outlook for their babies was even worse: maybe 5 per cent of children were stillborn and up to 10 per cent succumbed in the period immediately after birth. Even successful human childbirth was associated with a degree of pain that is probably unique among mammals. It is hardly surprising that childbirth has often been seen as the human race's penance for some ancient misdemeanour.

The challenge of birth

All this changed during the twentieth century. At last, humans had the opportunity to use that large head of theirs to solve some of the problems of childbirth. As a result, there have been spectacular declines in infant and maternal mortality in some countries, with decreases of perhaps 90 per cent between 1950 and 1990 alone. Although modern medical innovation has skewed the statistics slightly – by turning some stillbirths into deaths of newborn babies, for example – the general trend has been very much one of improvement. Yet one of the most shameful facts of modern life is that increases in maternal and infant survival have been far from universal. Although the improvements have largely been restricted to people in a few 'developed' countries, one of the main lessons of the history of obstetrics is that sheer technological wizardry does not improve the outlook for newborn babies as much as a few simple changes to birthing practice – teaching women how to tell when their birth is running into problems, or training local midwives to refer difficult births to hospitals. Although still not applied to many labours, most of these changes are easy and relatively inexpensive. It really does seem that, in the world as a whole, obstetrical disasters are now often avoidable.

Perinatal mortality is obviously a pressing concern for the baby. Even though she has established a successful pregnancy and developed normally, she still has a major obstacle ahead. Since those far-off days of early pregnancy when the intricacies of her body were pieced together, she has spent much of her time simply growing. As her size has increased, her placenta has kept pace with her needs, but soon everything will have to change. Birth is essential for her continued well-being, but it will be the most stressful experience of her entire life. Unfortunately for her, her chances of success during and immediately after her birth are not under her own control. She has been the arch-control freak for the last nine months, enthusiastically directing the progress of her own gestation. Now, however, she must enter the unpredictable and dangerous world in which the rest of us live.

Until recently, the most important requirement for successful birth and early infant life was a simple one: maternal nutrition. In many parts of the world this remains the case, and this is why poverty is still the most important world-wide cause of perinatal death. Three simple, but crucial things depend on the diet of a mother-to-be. These are the birthweight of her baby, her ability to undergo a successful, rapid labour and her ability to recover and supply sufficient milk to sustain her child. All three of these factors are obviously important for the baby's survival, which is why maternal diet is so important for a baby to thrive. In most parts of the world, the problem is a simple lack of energy or protein, although, in certain geographical regions, lack of individual 'trace' nutrients can have an equally dramatic effect on infant survival. Iodine, for example, is especially scarce in many mountainous regions, and in one recent study in Xinjiang, supplementation of irrigation water with iodine reduced infant and neonatal mortality by half. (I do not know if the link between low iodine levels and mountains is the reason why 'cretin' is a Swiss word.)

Like all medical statistics, the links between a mother's diet and her baby's chances of survival must be seen in context. Although diet is important, pregnancy is a fairly resilient process and can withstand considerable variations in maternal nutrition. When scientists link neonatal death to maternal diet, they are usually referring not to the erratic eating habits of pregnant women in the West, but rather to the long-term deficits in energy, protein and vitamins accumulated by women in poverty-stricken regions of the world. Indeed, the fact that diet is so important for babies makes it even harder to explain why morning sickness drives so many women effectively to starve themselves during pregnancy.

After malnutrition, the most important cause of perinatal mortality is probably infection. Immediately after birth, both mother and baby run an increased risk of bacterial infection; the mother because of the trauma of the birth process and the baby because of its immunological naïvety. Many postnatal infections are caused by rather mundane organisms that most people can easily tolerate at other times. To a newborn baby, however, these everyday bacteria must seem like a legion of mysterious new enemies waiting to attack. Of course, healthy babies can and do triumph over such adversity, but it takes little to shift the balance in the bacteria's favour. Babies weakened by maternal malnutrition or overwhelmed by the bacterial assault forced upon them by squalid living conditions can easily succumb to infections, usually in the form of diarrhea.

Mothers are also at risk of dangerous infections around the time of birth. Although the uterus seems to be fairly resilient to attack by most of the bacteria that live on women's skin, until recently the medical profession itself was a major source of life-threatening infection. Before the nature of infection was fully understood, obstetricians frequently spread fatal bacterial infections to new mothers simply because they did not wash their hands between patients. The reluctance of the medical community to accept its role in maternal death was shameful, especially as doctors continued to spread infection in this way long after the nature of bacterial infection was widely accepted. In developing countries today, many mothers are still dying as a result of simple lapses of hygiene. A trial in Malawi in which a few cents worth of antiseptic was used to wash the vagina during delivery, and to wash the newborn baby, reduced the number of fatal neonatal infections by two-thirds. It is sobering to realise that at the start of the twenty-first century, malnutrition and filth still carry off most mothers and babies who die around the time of birth.

As well as all the 'mundane' bacteria that attack newborn babies and their mothers, there are also some bacteria that specialise in causing miscarriage, prematurity or delivery of weak babies. Even in 'cleaner' parts of the world, some of these are still very real risks, although they are often fairly easy to avoid. One such bug is *Listeria monocytogenes*, mainly infamous for causing food poisoning. *Listeria* has a rather nasty habit of hunting out and infesting the placenta, and some obstetricians believe that many unattributed stillbirths are actually caused by *Listeria* infection. Yet *Listeria* is an abundant organism, and one to which we are all frequently exposed. Unfortunately, it has an unusual ability to grow at low temperatures, and so refrigeration can actually help it multiply up to dangerous numbers. Because of this, the main source of infection for pregnant women is contaminated dairy products, and soft cheeses in particular should be avoided during pregnancy. *Listeria* also causes miscarriages in sheep, so pregnant women should also avoid working with lambing ewes.

The main sources of another pregnancy bug, *Toxoplasma gondii*, are cat feces and undercooked marsupials. Avoiding the cat's litter tray and kangaroo steaks may sound like strange advice, but many pregnant women can and do catch this bug from exactly these sources.

The main lesson to be learned from the two great scourges of mother and baby – malnutrition and infection – is that simple things can make all the difference. In developing countries today, as well as in 'developed' countries until quite recently, much loss of life could have been avoided by simple preventative measures. Yet, as modern medicine becomes more and more technological, many people feel that simple preventative medicine is being neglected. It is not just that basic management of pregnancy has lost out to the rush for hi-tech childbearing in the struggle for funds, but also preventative obstetrics has an image problem — it is not as exciting or glamorous as heroically battling to save sick babies. Although this is a problem of preventative medicine in general, and not just preventative obstetrics, the problem is clearest when tiny babies are involved. In a world of limited medical resources, where do you direct your attention? Do you save more lives with cheap preventative medicine, or do you save fewer lives with expensive intensive management of sickly babies? The problem is that a sickly baby is a real, identified life in front of us, whereas the babies that could be saved by preventative measures do not yet exist, although they surely will. And what if that sickly baby is yours and those unidentified 'statistical' babies are someone else's?

This constant dilemma between improving the life of many potential babies and a few very real ones lies at the heart of the problem of managing pregnancy. Given the choice, many people would find it very hard to prioritise the welfare of 'potentially' ill infants above the welfare of 'actually' ill ones. What we are left with in our imperfect world is that resources get directed to rescuing sickly babies rather than preventing babies being born ill. Of course, this is no different in principle from redirecting funds from health education to heart transplants — it is just more poignant. Although more and more resources can be channelled into preventative pregnancy management in some countries, in others these resources simply do not exist. Then the choices faced by medical staff are more stark. In a poverty-stricken country, doctors will probably be viewed more favourably if they can dramatically rescue weak babies rather than lower stillbirth and infant death rates.

Most of the causes of perinatal death that are left once malnutrition and infection have been accounted for can be thought of as 'socioeconomic'. They include parental wealth, education and marital status, fetal sex and family size. Interpreting these factors can be difficult because different socioeconomic factors are often bound up together. It can be hard to tell when you have identified a cause of perinatal death rather than just something that causes or is caused by another factor that causes perinatal death. For example, illiteracy, malnutrition and smoking are all frequently claimed to be linked to infant death, but they are often also associated with each other. So which of them is the root cause that should be tackled?

Many doctors spend a lot of their time trying to dissect out all these interlinked factors. By carefully selecting groups for comparison, they can home in on the root cause or causes of medical problems. Their findings read like statistical summaries of all the sad little stories going on around the world. For example, infants seem to be more likely to die if they are born into large families, especially if they are later additions to the brood. Also, according to a Norwegian study, infant mortality is perhaps 50 per cent more common in babies of single mothers, even when their economic circumstances are taken into account. It is perhaps not surprising that maternal illiteracy has been linked to infant death, but the results of a study in Nicaragua showed that the babies at greatest risk are those whose mother is illiterate, but whose father is literate – the authors of this study questioned whether maternal empowerment was the factor controlling neonatal survival. Perhaps most tragically of all, one study in Tamil Nadu in India showed that up to a third of all perinatal infant death is due to deliberate neglect of female babies because male offspring are considered more valuable. Girls born to women who already had sons were 1.8 times more likely to die soon after birth than boys, but girls born to women with no living sons were 15.4 times more likely to die.

Medical care is also an important factor. The postnatal hospital care afforded to mothers and babies has been a controversial issue in recent decades, as postnatal hospital stays have become shorter and shorter in many countries. Certainly, the drive for mothers to become mobile again soon after birth is a great improvement on the system that prevailed until recently in much of Europe. It was once traditional for new mothers to spend a month or so 'lying in' after birth. This was thought necessary partly to give the woman time to recuperate after the toil of labour, but also to allow a time for communal thanksgiving for her safe delivery from her ordeal. It is now thought that this rest period was almost certainly counter-productive and even dangerous. The long period of recumbent inactivity caused many illnesses, especially venous thrombosis, in which blood clots form in the veins. Thrombosis was often fatal because these clots often became dislodged, drifted off in the bloodstream and blocked the large arteries that carry blood to the lungs.

So what is the ideal length of time for hospitalisation of new mothers? This question has no simple answer, as it depends on the medical care available after they leave hospital. Where adequate care and monitoring take place outside hospital, perhaps the minimum hospital stay should be the time after which emergencies requiring immediate medical attention are unlikely. After all, the big advantage of giving birth in a hospital is having expert staff on hand if something goes badly wrong. The time that elapses before the risk of sudden unheralded emergencies is past is a matter of considerable debate, but eighteen hours is probably a reasonable figure. If something is going to go rapidly, badly wrong, it will probably have done so by then.

Hospitalisation is, of course, a very controversial part of the modern birth process. Many women are now deciding to have their babies at home, away from the impersonal hospital environ-ment. Yet this has led to a long-running argument between those who believe that home is best and those who claim that hospital is safer. It does seem strange that, at a time when most medical services in developed countries are becoming ever more centra-lised into large hospitals, so many women are eschewing hospital birth. At times the argument has become acrimonious, with one side claiming that hospital is the best environment to deal with the unpredictable dangers of birth, and the other side pointing to statistics that often show better success rates for home births.

The problem really boils down to the question of comparing like with like: high-risk births tend to take place in hospitals, whereas obstetricians and midwives are more willing to allow lower-risk deliveries to take place at home. The pro-hospital camp believe that this explains the high success rate of home births and the pro-home lobby claim that their surveys have taken all the extra risk factors into account. So the home versus hospital debate remains unresolved. I cannot, however, believe that home is a safer place to have a baby. Some babies are saved by emergency Caesarean sections precipitated by unforeseeable circumstances, and some mothers are saved from bleeding to death from unexpectedly rapid haemorrhage by immediate blood transfusions. Thankfully, these cases are rare, but they do occur and they can happen to any mother, be they 'high-risk' or 'low-risk'.

There is one more cause of perinatal death, whose mention will no doubt raise a collective groan among those prospective parents affected. Yet the fact remains that this cause of perinatal problems is the most preventable of all: smoking. If ever there was a good reason to give up smoking then pregnancy is it, and that applies to fathers too. It is hard to estimate the pregnancy problems caused by smoking, but it is probably fair to say that in many countries, tobacco has replaced poverty as the main cause of unsuccessful pregnancies. Smoking has important, indiscriminate effects on developing babies. Before birth, smoking can reduce the amount of oxygen reaching the fetus, can retard fetal growth and may even cause portions of the placenta to degenerate. One study in Missouri showed that women who smoked were over 80 per cent more likely to produce low birthweight (less than 2500 grams or 5.5 pounds) babies than women who did not. Because of this, smoking causes fetal deaths and reduced birthweight, but babies are not safe from smoking even after birth. Although no one knows why, smoking is also strongly linked to 'cot death' during the first year of life.

Clearly, the whole tenor of pregnancy changes dramatically as birth approaches. Instead of being in control of events, our baby becomes more dependent on her mother. She can only hope that her benefactor eats well and avoids soft cheese, cat droppings, kangaroo-burgers and tobacco. No longer a serene little being drifting in her personal jacuzzi, the baby will soon have to deal with the messy, unpredictable outside world, just like the rest of us.

Before she even gets there, she and her mother must survive the birth process itself, which is one of the most perilous ordeals either of them will ever endure. The birth process is complicated and committing – if a single stage fails, all is lost. But why is birth so difficult?

The great escape

All good things must come to an end, and for the baby pregnancy really has been a good thing. Excellent lodgings, individually prepared food and a congenial atmosphere have all made for a rather pleasant stay, but now the time has come for the little visitor to leave.

The uniqueness of the process of birth is deeply ingrained into the human psyche. Birth has been the subject of some of the most unlikely myths and hoaxes – it is such a powerful process that people seem to be able to believe that it can do almost anything. One of the most extreme birth myths was the miraculous labour of Countess Margaret of Henneberg on Good Friday 1276. The young countess certainly died on this day, but only later was it claimed that she perished immediately after giving birth to 365 live mouse-sized children, who also later died. Not surprisingly, the site of this prodigious labour, at Loosduinen in Holland, became a place of pilgrimage for barren mothers. We can only speculate about the origins of the myth, but the story chimes with the modern finding that one of women's greatest fears of birth is that they will in some way 'fail' to be 'good' mothers. The story may have become attached to the countess because she had almost completely retired into a nunnery after the death of a previous son.

In most myths there is a grain of truth, and there exists an animal equivalent of Countess Margaret of Henneberg. As if to prove that animals can exploit any available method of reproduction, there is a South American rodent called a plains viscacha that operates the same system as the countess. Most female animals produce roughly as many eggs as they produce children – women usually make one a month, sows make around fifteen per cycle – but the viscacha has different ideas. Every cycle, a female viscacha ovulates a few thousand eggs, and many of the eggs are fertilised if she is mated. The reason that the world is not knee deep in viscachas is that most of the embryos die early in pregnancy, and viscachas actually produce litters of only two or three.

One of the greatest birth hoaxers of history was Mary Toft of Surrey, and for some time she successfully fooled most of the London aristocracy and much of the medical profession. In April 1726 she was working in the fields near her home in Godalming when she saw a rabbit, which she chased to no avail. During the next few nights she dreamed of eating succulent rabbit meat and during the day, so the story goes, she was obsessed with rabbit meat. A few months later she miscarried what appeared to be a mass of flesh, and in subsequent days she continued to miscarry what were later identified as the internal organs of pigs. Her theatrical masterstroke came when she apparently started to miscarry dead rabbits, sometimes skinned and sometimes cut into neat portions. This continued, and her doctors were keen to bring her to the attention of the London establishment. She soon became a celebrity in the capital, although opinion was divided as to the cause of her complaint. Mary's cover was blown later in the year when a servant claimed she had bribed him to bring her a dead rabbit. She then confessed all – that she had developed the ability to retain pieces of rabbit within her vagina, so she could then 'miscarry' them some hours later. The belief that babies could be affected by 'maternal impressions' – that they could take on the form of animals that their mothers saw while pregnant – was widespread at the time, and probably explains why Mary managed to hoodwink the medical 'experts' for several months.

Birth may retain such a hold over the human imagination because it often strikes fear into prospective parents. Fear of birth – variously called tocophobia, maleusiophobia, parturiphobia or lockiophobia – is one of the commonest human fears. Midwives spend a great deal of their time trying to allay the fears that parents have of this unavoidable and painful experience. Fear of birth also has a great effect on the outcome of labour – a group of Swedish researchers found that women with significant fear of birth are three times more likely to have their baby delivered by Caesarean section, and a different study found that they are also more likely to suffer emotional problems after birth. Also, pain is a strange phenomenon, and it has a large psychological component – women who are more confident about their ability to undergo a normal labour experience less pain during childbirth.

A degree of trepidation before birth may be justified, but women and men differ greatly in what they actually worry about. Women worry more about the baby. They have been reported to worry most about, in descending order: fetal malformation or injuries; the possibility of dramatic medical intervention; the strange environment of the hospital; themselves doing something wrong; uncertainty about how the child will be delivered. Men seem to worry more about their partners than their child – they worry about their partner's suffering; the possibility of dramatic medical intervention; fetal malformation or injuries; their own powerlessness; their partner's death. One notable exception to the mother's list of fears is pain, which features more highly on the list of the partner, who does not actually have to experience any pain himself. Unlike all the worries on their list, perhaps women see the pain of birth as something that no one can prevent, so they do not think to mention it. Also, it is tempting to suggest that they fear for the baby's welfare because they are the ones that have felt it growing and kicking inside them.

A surprising number of men show physical signs of empathy with their pregnant wives. This sympathetic pregnancy is called couvade (from couver, the French for 'to hatch'). Often starting in the third month, between 10 and 65 per cent of fathers have been claimed to report some of the signs of couvade: depression, indigestion, nausea, anorexia, food cravings, weight gain, diarrhea or constipation, insomnia, nose bleeds and even abdominal swelling. Many fathers report an improved sense of well-being during the second trimester, but increasing feelings of depression during the third – in exact synchrony with their wives. A more severe form, psychotic couvade, has also been reported. Couvade has been claimed to be a sign of anxiety, doubts about the paternity of the child and a subconscious attempt to empathise with the fetus, although I suspect many women would call it unhelpful whining.

Yet there does seem to be a hormonal basis for couvade. Levels of prolactin in some fathers' blood increase by 20 per cent and levels of the steroid hormone cortisol can double. Couvade also leads to a little-known behavioural effect that few fathers would consider faking – after birth, couvade sufferers are more likely to cradle the baby's head in their right arm (80 per cent of parents cradle it on the left). Couvade has become 'ritualised' in many cultures, and almost ceremonial forms of couvade were reported on the travels of Strabo, the ancient Greek geographer, as well as by Marco Polo. Basque men, for example, may take to their beds during childbirth and simulate labour pains. It has been suggested that couvade is a public way for men to claim paternity, and thus possession, of a new child. Alternatively, it may just be a conveniently institution-alised way to avoid being present at the birth.

For many parents-to-be, one of the aspects of pregnancy that increases their anxiety is that they do not know when birth is going to start. Mothers often feel powerless – the time of birth seems completely out of their hands. The timing of birth is obviously very important for the fetus too – in 'developed' countries, perhaps 75 per cent of perinatal deaths occur in the 10 per cent of babies that are born prematurely. Premature labour can start as early as twenty weeks, and many doctors believe it is often caused by infections of the uterus or maternal stress. Yet the causes of early labour are surprisingly poorly understood, as indeed are the factors that control the length of normal pregnancies. For centuries, scientists and doctors have wondered who controls the onset of birth – whether the baby or the mother is in control. Does a baby voluntarily check out of its hotel in good order, or must it be evicted?

For a long time, the timing of birth was one of the most mysterious aspects of pregnancy. Despite this, it was obvious that labour does start with remarkable consistency at the appointed time, so there must be something that controls when it begins. For centuries it was thought that the birth supervened when the placenta could no longer cope with the demands of the growing baby. In the fifth century BC, Hippocrates stated that 'When the infant has grown large and the mother can no longer feed him, he struggles and breaks out into the world, free from all bonds.' After Hippocrates' time, the fetus was thought to become progressively more restricted by the limitations placed upon it as pregnancy went on. Birth was the baby's way of 'baling out' before things got any worse. This idea of birth as a response to creeping fetal deterioration was even compared to the rigours experienced by mountaineers. In 1946, the British physiologist Henry Barcroft coined the term 'Everest *in utero*' to describe the small amounts of oxygen available to the fetus.

This idea of the baby simply 'giving up' when food and oxygen are running low has flaws, however. First of all, it seems rather risky as it allows the baby little margin of error. If a baby's placenta is very slightly underdeveloped, this miscalculation might mean that the baby has to be born prematurely. This seems rather a harsh penalty to pay for a minor developmental mistake. Also, it would be strange for mammals to evolve a system that progressively starves and asphyxiates their offspring before they are born. Finally and most importantly, there is simply no real evidence that things actually do get worse for the baby just before birth. Babies do not have to be 'starved out'– after all, the eventual aim of a fetus is to be born, not to stay where it is. Instead of being forced out by some sort of placental food embargo, all the baby needs is the right signal to tell her that the time is right.

The past ten years have seen great strides in our attempts to find out what this signal is. Previously, the inability to do experiments on women at the end of pregnancy made scientific progress very slow. We were able to untangle the mechanisms controlling birth in other mammals, but what we discovered did not seem likely to explain the start of labour in people. More recently, however, new techniques that allow us to study the molecules made by the placenta have allowed us to resolve the question of whether it is the mother or the baby who controls birth.

We have learnt about the initiation of birth by taking our understanding of birth contractions and 'working backwards' to find out what sets them off. Because obstetricians sometimes need to deliberately induce birth for medical reasons, scientists have always had an ethical justification for studying what causes contractions in women. Fortunately, labour contractions have turned out to be one of the few parts of the birth process that are fairly similar in different mammalian species. So contractions are a good place to start our quest for the switch that turns on birth.

When obstetricians want to start off birth contractions, they usually use prostaglandin. This is the chemical that drives the actual muscular contractions of the uterus as well as the softening of the cervix that allows it to open to let the baby through. Obstetricians sometimes find that, for some reason, prostaglandin does not work on its own, and so they administer oxytocin as well. This combination is usually effective at getting contractions under way.

During normal birth, the prostaglandin probably comes from within the uterus, helped along by oxytocin from the pituitary gland. Oxytocin and prostaglandin also work together to destroy the corpus luteum in red deer – here the same two hormones are now directing the process of birth. It is almost as if the roles of the hormones have been stretched in time as mammals have evolved longer and longer pregnancies. Put simply, their action is to end pregnancy. If their attempt to destroy the corpus luteum is foiled at the start of pregnancy, then they act to precipitate birth at the very end.

The anti-pregnancy alliance of oxytocin and prostaglandin raises an important question. If oxytocin and prostaglandin are such potent pregnancy-terminators, how have they been kept in check all the way through the last nine months? Their activity must have been carefully controlled or pregnancy could have been cut disastrously short. When we look at how oxytocin and prostaglandin are restrained, we will also be going one step back in the chain of events that leads to labour. We are edging back in time to that holy grail of pregnancy biology – the trigger to birth.

At the end of pregnancy, just as at the start, oxytocin and prostaglandin are under the control of the steroid hormones estrogen and progesterone. Unfortunately for us, this is where different mammalian species start to do things differently. In some species, most steroids are made by the mother's ovaries, just as they were at the start of pregnancy. However, in other species, humans included, the placenta has long since taken over as the main source of steroids. Because of this, sheep have proved to be a useful model for human pregnancy because their placenta also produces most of their steroids.

Around the time of lambing, changes are afoot in the way that the sheep placenta makes its steroids. There seems to be a shift away from making progesterone and instead the placenta makes more estrogen. So, as lambing aproaches, estrogen slowly starts to predominate over progesterone. Estrogen probably starts the birth process because it makes the maternal pituitary produce oxytocin and the uterus secrete prostaglandin. So the shift towards estrogen predominance is probably instrumental in activating birth contractions in sheep. Our woolly friends are being extremely helpful.

But if estrogen is so important, why does the placenta start making more of it? What, or who, tells it to? As we discover each link in the chain, it seems we have to look ever earlier in pregnancy to find the link that precedes it.

In sheep at least, the end is in sight. The trigger that induces the placenta to make more estrogen is a burst of production of cortisol by the lamb's adrenal glands. The adrenal glands are two small but vital blobs sitting just in front of the kidneys and they make a variety of hormones. Cortisol is a steroid, but not a 'sex' steroid like progesterone, estrogen or androgens. It is completely different – related instead to hydrocortisone and dexamethasone, the steroids that doctors use as anti-inflammatory drugs. A sudden surge of fetal cortisol is the trigger that starts the lamb's placenta turning progesterone into estrogen.

Just like the ovarian steroids we looked at in Chapter 2, adrenal cortisol secretion is controlled by the pituitary gland. The pituitary controls the adrenal with another hormone called ACTH, which stimulates cortisol release. In turn, pituitary production of ACTH is under the control of a brain hormone, CRH. By reaching fetal CRH we have finally chopped through all the hormonal undergrowth obscuring our final answer: lambs initiate their own birth because their brains make CRH when they are ready to be born.

Fetal brain releases CRH

↓

Fetal pituitary releases ACTH

↓

Fetal adrenal releases cortisol

↓

Placenta and uterus make more estrogen,
less progesterone

↓

Estrogen dominance

↓

Prostaglandin induces birth

So lambs, and not ewes, control the timing of birth. In a final attempt to control their destiny, lambs choose the occasion of their entry into the world. Birth is the last gasp of the fetal control freak – in sheep, at least.

When scientists first tried to extend the work they had done with sheep to humans, they ran up against some problems. Although they found evidence of estrogen dominance around the time of birth, injecting women with cortisol did not seem to induce labour. Also, even though babies churn out extra cortisol as birth approaches, it does not appear to be essential for labour to start – babies that cannot make cortisol are not born late, as you might expect if cortisol were the trigger to labour.

Human babies produce a different hormone in their adrenal glands to kick-start birth, and it has yet another jargon name, DHEA. This makes its way to the placenta, where it is the raw material from which the placenta manufactures estrogen. So the human baby's adrenal gland is important in starting birth, but it produces DHEA instead of cortisol. Bringing the story back closer to that of sheep, the human fetal adrenal is induced to make DHEA by ACTH from the baby's pituitary, and, again like lambs, the ACTH release is driven by CRH.

The similarities between humans and sheep now fail us at the last minute. Unexpectedly, the CRH that drives the human birth process comes not from the baby's brain, but from the placenta.

Placenta releases CRH

↓

Fetal pituitary releases ACTH

↓

Fetal adrenal releases DHEA

↓

Placenta makes estrogen from DHEA

↓

Estrogen dominance

↓

Prostaglandin induces birth

So it seems that the timing of birth in humans is not under the control of mother or baby. The CRH signal comes not from a brain, an organ that one might expect to be able to keep track of the progress of pregnancy, but from the placenta: the human placenta triggers birth.

The placental instigation of birth in humans raises some important issues about pregnancy. First, we do not know why mothers and babies find it more convenient to measure their pregnancies with their placenta, rather than with one of the two brains available – it is not at all clear why the placenta is better placed to keep track of pregnancy. Also, the discovery of placental CRH has led to new ideas about how the clock of pregnancy ticks. The placenta makes CRH from a very early stage of pregnancy, perhaps as early as twelve weeks after conception, and its CRH output increases steadily until it eventually triggers birth.

The climb of CRH is so inexorable that it is possible to predict the time of birth from CRH levels measured in the middle of pregnancy. From then on, the upward climb of CRH secretion can be predicted as far as the point at which it will initiate labour. It now seems that this creeping increase in placental CRH secretion is the timekeeper of human pregnancy. The placental stopwatch is started soon after the placenta is formed and thereafter it reliably counts down to zero hour. Probably, the length of human pregnancy is decided by how this stopwatch is set at the beginning of the countdown.

So the human baby has sacrificed its ability to control the time of its own emergence into the outside world. The human placenta has taken over the job of the brain in starting birth. In Chapter 2 we saw that the invading human placenta produces hCG, a hormone related to pituitary hormones, to start pregnancy. It is as if the placenta had 'stolen' a brain hormone to make hCG. Now we see that a similar process has evolved to control the end of human pregnancy as well. The placental 'brain' is in action again, stealing another brain hormone – CRH – to dictate the timing of birth.

The trigger to birth sets in train a series of events that leads inexorably to the expulsion of the baby from its uterine idyll. Although sometimes preceded by the 'show' (the loss of a plug of mucus and blood from the vagina), or even the 'waters breaking' (the rupture of the trophoblast and amnion membranes to allow the amniotic fluid to escape), contractions are often the first sign that labour has started. Some women experience minor 'tighten-ings' of the uterus, called Braxton Hicks' contractions, for many weeks before birth, and for them the onset of true labour can be rather indistinct. There is nothing indistinct about full-blown labour contractions, however. The first stage of labour is domi-nated by these, and dominated by the actions of prostaglandin.

By the time of birth, the uterus has become the largest muscle in the woman's body and it is prostaglandin that stirs this giant into action. The cells of the uterine muscle are interconnected, so they contract as a coordinated unit: ripples of contraction undulate through the uterus and pull the lower part of the uterus up from the pelvis to straighten the birth canal. Yet the baby still cannot see 'the light at the end of the tunnel' because the cervix, the outlet of the uterus into the vagina, is closed. The cervix acts as a muscular draw-string closing the outlet of the uterus, but it starts to relax and dilate as prostaglandin gets to work on it. In some species, such as guinea pigs, special hormones even loosen the joints between the different parts of the pelvis so that the whole pelvis breaks open to widen the baby's escape hatch. This relaxing of the pelvis also occurs to a lesser extent in women.

By now, the contractions of the first stage of labour have pushed the baby's head against the cervix. Nerves in the cervix course back to the mother's brain, where they stimulate the secretion of more oxytocin. This then increases the amount of prostaglandin acting on the uterus. The more pushing, the more oxytocin: the more oxytocin, the more pushing. The first stage of labour takes an average of twelve hours in women's first labour, and by the time it is over, there can be no going back.

Most babies now slide down the sculpted sides of their mother's pelvis so that they end up head down, with their belly facing her spine and their arms at their sides. Most babies do not directly face their mother's spine, but are slightly tilted to the left or right – for some reason, more babies face their mother's right side than the left. This is the commonest position in which human babies enter the birth canal. Yet most other mammalian infants pass through the pelvis the other way round, with their heads first, but with their back facing their mother's spine and with their forelegs extended in front of their head, like a springboard diver. Because of this, foals and calves describe the arc of a diver as they are thrust upwards into the pelvis, pass through it and fall out with their forehooves hitting the ground first. Most animals use this spine-to-spine system and it is not obvious why human babies have to face their mother's spine. The reason for humans' unusual birth position may have something to do with the shape of women's pelvises now that they have been adapted for walking upright. More likely, the unusual birth posture is all to do with which parts of the baby are widest. Any mother will tell you that the human baby's head is the sticking point, whereas in many other mammals the shoulders or hips are the widest part.

The second stage of labour is the actual expulsion of the baby. The mother now augments her uterine contractions with strong contractions of her abdominal muscles. Although abdominal contractions start spontaneously, they are much more under the woman's control than uterine contractions, even to the point where she can stop them when she wishes, or at least when she is told. Many women say that the second stage of labour is actually less painful than the first, apparently unproductive stage. They often say that this stage feels different – the sharp and piercing pain of the first stage is replaced by the feeling that their pelvis is crammed full of an object that must be passed. The pain may also recede slightly because the mother now knows that she has a degree of control, as well as the knowledge that her baby will soon be born. Many women naturally experiment with different birth positions during the course of a single labour. Although many midwives and obstetricians no longer feel strongly about the ideal position for a birthing mother, the 'traditional' supine position has become less popular recently, perhaps because it requires the mother to push the baby up as well as out.

As her mother's contractions ebb and flow, the baby is rocked back and forth in the birth canal, each time creeping slightly closer to freedom. This is a stressful time for both mother and child. The mother's blood pressure can become extremely high, and each contraction also slows the baby's heart noticeably. These heart slowings – or 'dips' as they appear on fetal heart traces – are useful indicators of the baby's well-being. Dips themselves are normal, as long as they do not persist after the contraction is finished. So if you hear a midwife referring to a baby's dips, this does not mean that anything is wrong. In fact, careful monitoring of dips during the second stage is a good way to reduce the chance of things going dangerously awry.

The backwards and forwards rocking motion of the baby stops when her head finally appears at the vulva. This 'crowning' is an obvious sign that something exciting is about to happen. This is especially true because the human baby's head is not just the first bit – it is the worst bit. If a mother is lying on her back, then usually her baby will be facing the floor with the top of its head protruding from the vagina. Unfortunately, nature has bestowed upon women a birth canal with a bend in the middle. Because of this, the baby's head must now be tilted back to allow it to be squeezed through the lower part of the birth canal. The nape of the baby's neck is lying against the front of her mother's pelvis, forming the pivot about which the baby's head is turned. Now, the midwife often turns the baby's head so that her face pops out of the birth canal. This is one of the most important moments in the birth process, and it is usually the time when the midwife tells the mother to stop pushing. It is also one of the most painful moments and is often accompanied by what the obstetrics textbooks politely refer to as 'vocalisation.'

After this tilting procedure, things get slightly easier. Once the head is squeezed out, there is usually a slight pause as the shoulders negotiate their way into the outside world, and after that the whole process is soon over. For a first-time mother, the second stage of labour takes an average of one hour to run its course. Our baby has completed her terrible journey from the inner world to the outer. All she had to do was arrange herself into a suitable posture and prepare to suffer. Her mother not only had to suffer, but also had to do most of the work as well. All mammalian mothers put a lot of effort into birth, but there are good reasons to think that the pain of childbirth is especially severe for women. Women have been left to deal with the consequences of the unstoppable evolutionary drive for ever-larger human heads. The only help they have been given to fortify them is some pain-relieving endorphin hormones and improved blood clotting around the time of birth. When, as they usually do, these prove woefully inadequate, nature allows women a kind of selective amnesia that seems partially to erase the memory of childbirth. There are several studies in which women apparently failed to recall the pain they experienced during birth accurately – indeed, they usually underestimate it. This natural valium must work quite well – after all, lots of women go on to repeat the ordeal. There are also scattered reports of otherwise normal women during the second stage of labour entering a form of trance, in which they become completely unresponsive to the outside world.

Medicine has completely changed childbirth. As well as tackling the twin scourges of malnutrition and infection, modern practice has also dramatically reduced the chances of the obstetric disasters that used to fill graveyards around the world. Many of these happened because evolution has left babies little room for error when it comes to their escape. If they are slightly too large, or are in the wrong position, then the relentless unstoppability of birth often consigns them to disaster. Yet if women have their babies in hospitals with legions of trained midwives, obstetricians, paediatricians and blood transfusion experts, then these disasters are far less likely.

Of course, the main reason that the problem of baby-will-not-come-out-through-that-hole no longer leads to the disasters it once did is the advent of the Caesarean section. When babies and pelvises are incompatible, the only solution is to miss out the pelvis and remove the baby through its mother's abdominal wall. By completely avoiding the pelvis in this way, the Caesarean operation must have alleviated more human suffering than just about any other medical procedure. The Caesarean is also thought to be one of the oldest surgical operations and was attempted long before the advent of antiseptics or anaesthesia. Performing it under less-than-ideal conditions was, it seems, better than not performing it at all.

The origins of the Caesarean operation are obscure, as one might expect for a procedure so ancient. It may have originated as a way of delivering a baby when its mother was already moribund, if only to allow it a separate burial. In classical Rome, the operation was compulsory as mothers slipped towards death to ensure that production of new empire builders was not compromised by the inconvenient inadequacies of human childbirth. This may be how the 'Caesarean' got its name. The more commonly held belief that Julius Caesar was born in this way is unlikely as there are reports of his mother surviving to a ripe old age. From the classical era onwards, Caesarean section was carried out sporadically until the nineteenth century. The operation was often done in the mother's home by untrained staff, thankfully distant from the filth of medieval hospitals.

Although the resources were often available to nineteenth-century surgeons to carry out Caesarean sections, many of them still often demurred when the possibility was suggested. After all, was not childbirth a trial sent down by God himself as women's penance for their part in the fall from Eden? 'Unto the woman he said, I will greatly multiply thy sorrow and thy conception; in sorrow thou shalt bring forth children.' Perhaps it was easier for the more male-dominated medical profession of the nineteenth century to be restrained by such doubts.

Our baby has no time for such theological diversions. She has just landed in the cold, dry, unwelcoming world. No fluids to bathe her, no placenta to feed her: she is now forced to undergo the most dramatic transition she will ever face. Everything in her little body will have to change. Almost as inspiring as pregnancy itself is the fact that we ever learn to cope with life when pregnancy is over.

The world outside

Some babies do not have to breathe. Julia Creek is an impressively remote town in outback Queensland. When I was there in 1994, little did I realise that there was such a creature as a Julia Creek dunnart. It was not until 1999 that I learnt of this obscure mouse-like marsupial, which lives in the cracks in the sun-baked outback mud. A paper appeared in the scientific journal *Nature* making the remarkable claim that newborn dunnarts do not make any breathing movements. It seems that the baby Julia Creek dunnart is so tiny that it can absorb all the oxygen it needs directly through its paper-thin skin. Only as it grows does the infant dunnart start to inhale and exhale. This strange little creature is, of course, the exception that proves the rule. Newborn baby mammals face many challenges, but the most urgent is that they have to start breathing. Most of the other changes they will have to make can be quite gradual, but the first gasp of life must take place, and it must take place quickly.

We last looked at our baby's lungs in Chapter 3 when they formed as two outpouchings of the front end of the gut. These branched repeatedly, sending millions of tiny ramifications backwards into the chest. Each of these branchlets, a tiny thin walled sac with an abundant blood supply, eventually became an alveolus or air pocket in the mature lung. Before birth, however, the lung spaces are full of fluid and they have no role in gathering oxygen for the baby. That, as we have seen, is the job of the placenta. Yet, although they are redundant throughout pregnancy, the lungs are far from inactive. From as early as the twelfth week of pregnancy, the baby's chest starts to make erratic movements, and over the next twenty-eight weeks these fetal breathing movements become more regular and pronounced. Although the lungs are still filled with fluid, this does not stop the chest muscles doing some exercises in anticipation of birth.

As birth approaches, not only are the chest muscles preparing for their vital job, but essential chemical changes are also under way in the lungs themselves. These chemical changes are an attempt to overcome an embarrassing design fault in all mammalian lungs (a fault that superior beings, such as birds, have avoided). Lungs contain millions of tiny air chambers coated with an extremely thin layer of liquid to help them absorb oxygen. Unfortunately, when a gas-filled, liquid-lined air space is expanded, a force called surface tension acts to oppose the expansion – it is surface tension that pulls soap bubbles into taut little spheres. Even worse, surface tension is strongest when the air space is small, as it is in the air chambers in the lungs. As a result, our lungs must constantly fight against surface tension as they expand and contract. Mammals' solution to this problem is to add a detergent-like chemical called surfactant to their lung fluid. Surfactant reduces surface tension in our lungs so that we can force them to expand. Without surfactant, you would be dead.

In the last weeks of pregnancy, the baby has been desperately bringing her production of lung surfactant up to speed. She does not need it when she is in the womb because her fluid-filled lungs do not have to fight against surface tension, but having surfactant before birth does no harm and so babies make it in advance. Rather neatly, it appears that the stimulus to surfactant manufacture in the human lungs is cortisol – the hormone that initiates birth in sheep. This is why cortisol is present in increased quantities before birth in humans, but does not induce birth – its role in humans is to prepare the lungs for breathing air. It is hard to overstate the importance of the fetal adrenal gland: it produces the DHEA that allows the baby to be pushed into the outside world and it makes the cortisol that ensures she can breathe when she gets there.

Surfactant is extremely important in the management of premature babies. Many premature babies are born before they have a chance to make any surfactant and until recently these babies usually died. If doctors think premature birth is looming, they can give the mother a large dose of a cortisol-like drug to try to start surfactant production, but very often they do not get enough warning to be able to do this. Instead, many of the recent improvements in survival rates of premature babies have resulted from an understanding of how babies' lungs work when they do not have enough surfactant. (Incidentally, the use of drugs to induce lung maturation is the beneficial side of the capriciousness of human reproduction: if you inject a pregnant ewe with cortisol, you will precipitate premature birth. Because human fetuses use different hormones to induce birth and lung maturation, obstetricians can help lung maturation without causing labour.)

Underdeveloped lungs are the premature baby's main liability, but several important advances have led to the large-scale rescue of babies that would otherwise have died because of breathing difficulties. One approach has been to supplement the baby's lungs with artificial surfactant once it has been born. However, it has been difficult to find the best way to get the artificial surfactant in: the ideal method would mimic the natural production of surfactant by mature lung cells. An alternative approach has been to accept the shortcomings of the immature lungs and simply to avoid expanding them against the force of surface tension. Remarkably, it seems to be possible to oxygenate premature babies without the large, slow inhalations and exhalations of normal breathing. Sufficient oxygen can be instilled into a baby's lungs by frequent bursts of air, or even by introducing carefully directed continuous jets of air. Most ingeniously of all, the surface-tension problems of immature lungs may be overcome in the future by making babies breathe liquid. Studies have clearly shown that rodents can breathe successfully when submerged in special liquids saturated with oxygen. Maybe immersing premature babies in a warm tank full of dissolved oxygen will prove to be the best way to ventilate them.

All this need not concern our baby, however. She has been lucky. Because she was not born too early, she has generated a healthy surge of adrenal cortisol and her squashed lungs are liberally soaked in surfactant. Her problems are not yet over, of course. She has not yet breathed and her lungs are still full of fluid. That first breath is all-important – if she can make one breath, then all the rest should follow.

There are probably two main things that trigger the first crucial gasp. The first is a passive mechanism and is probably not essential for many babies. When the baby is extruded through the birth canal, her chest is squashed as never before. As she flops out into the world, her elastic chest springs back to its original shape. It has been suggested that this simple elastic recoil passively sucks a gasp of air into the lungs. Of course, this springing back into shape cannot be essential for breathing to start, because babies born by Caesarean section are not crushed in the birth canal and most of them breathe perfectly well. Also, babies born into warm water consistently manage to delay their first breath until they reach the surface.

Probably more important are babies' active attempts at gasping. Babies are often partially asphyxiated by the time they are born and asphyxia is a powerful trigger of gasping – try holding your breath under water for a minute and see what happens when you surface. Other things that babies use as cues for gasping are cold and exposure to the air. This is why babies born in warm birthing pools do not breathe until they reach the surface. Cold is an excellent stimulator of breathing: the Russian artist Marc Chagall was thought to be dead at birth, but revived when he was dropped into a bucket of iced water (and this unusual revival method was a theme that recurred throughout his career in his dreamlike pictures). It may sound cruel, but pain is also one of the best ways to start neonatal breathing – hence the traditional slap once administered to newborn babies. I have successfully revived many newborn mammals by pinching ears, squashing toes and nipping noses.

Once the first gasp has been taken, the baby's battle is nearly won. The gasp draws air into the lungs and does most of the initial work to overcome surface tension. Other breaths follow much more easily, but the ability of the lungs to absorb oxygen remains poor until all excess fluid is removed. Unaided, most mammals' lungs can clear this fluid in a matter of minutes by an efficient drainage system that channels it into the blood. If superfluous lung fluid is still a problem, then it can be removed artificially. Paediatricians and vets differ only in the refinement of their approach. Paediatricians suck the fluid out with a tube, whereas vets drain it by twirling the infant around by its back legs or hanging it over a barn door.

The first gasp sets off a series of changes in the baby's circulation that will adapt her life support system for survival in the outside world. She now has an extensive plumbing job on her hands. While in the womb, her circulation was designed to receive oxygen and food from the placenta and distribute them around the body. Now, suddenly, it must collect oxygen from her lungs and food from her intestines instead. As soon as the first breath is taken, parts of her circulatory pipework are dramatically altered to divert blood flow to where it is needed, and other changes follow in the next few days. It is astounding that such a profound redirection of the circulation is possible without any apparent adverse effects on the baby.

Figure 6 shows the pattern of the circulation after birth, and superimposed on the picture (in bold) are the differences present before birth. As William Harvey revealed in *De Motu Cordis*, after birth, the blood will circulate successively through the two distinct sides of the circulation, the right (lung) side and the left (body) side. Try following the arrows in the picture. Starting in the right side of the heart, blood is pumped through the right atrium and ventricle into the pulmonary arteries and thence to the lungs. There, the blood is loaded with oxygen before flowing back to the heart in the pulmonary veins. These empty into the left side of the heart, which then pumps the oxygenated blood to the rest of the body through the aorta. Once the body has extracted all the oxygen it needs, the blood returns via the vena cava to our starting point: the right side of the heart.

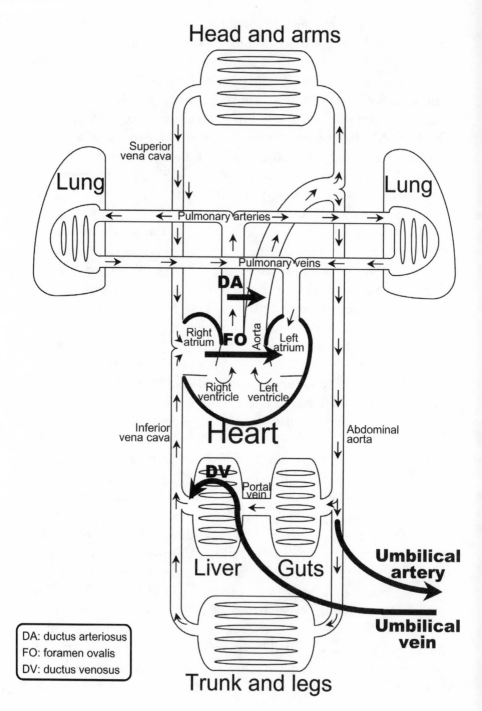

Head and arms

Superior vena cava

Lung

Lung

← Pulmonary arteries →

Pulmonary veins ←

DA

Right atrium

FO

Aorta

Left atrium

Right ventricle

Left ventricle

Inferior vena cava

Heart

Abdominal aorta

DV

Portal vein

Liver

Guts

Umbilical artery

Umbilical vein

DA: ductus arteriosus
FO: foramen ovalis
DV: ductus venosus

Trunk and legs

Several things are done differently in the fetus. She has no need to send much blood to her lungs because she does not collect any oxygen there – so blood is diverted away from the right (lung) side of the heart into the left (body) side through two holes. One hole is the foramen ovalis ('oval hole'), which allows blood to leak from the right atrium into the left atrium. The other is the ductus arteriosus, which diverts blood from the pulmonary arteries to the aorta. This right–to–left shunt almost entirely bypasses the lungs, preventing the useless pumping of blood through these as yet redundant organs.

After birth, this arrangement has to change. The right–to–left shunts in the heart must close off so that all blood draining from the body is sent to the lungs for reoxygenation. When the baby takes her first gasp, the blood pressure in her right atrium falls and a valve snaps shut over the foramen ovalis, closing off the escape of blood. Over the next few days, the ductus arteriosus also closes off, completing the separation of the right and left sides of the heart. By an elegantly simple system of valves, the baby has hastily switched on its lung circulation. Just how essential these changes are is made clear when they fail to take place. Some of the commonest heart problems in children result from failure of one of these two holes to close. Fortunately, many heart problems in babies are curable by surgery, mainly because they are of this simple 'plumbing' variety.

Lower down in the body are other special features of the fetal circulation that have to be removed at birth. First of all, the enormous umbilical vessels course out of the navel towards the placenta to gather food and oxygen and offload waste. The

Figure 6. The circulation of the blood, and how it differs in the fetus. In adults, blood is pumped around the body by the left side of the heart. The blood then returns to the right side of the heart in the veins. The right side then pumps it into the lungs to absorb oxygen and from there it drains back into the left side.

In the fetus, the lungs are effectively bypassed by leakage of blood through two holes in the heart: the foramen ovalis and ductus arteriosus. Also, much blood is pumped through the umbilical artery to the placenta, where it absorbs oxygen and nutrients. This blood returns to the baby in the umbilical vein, which bypasses the intestines and liver by means of a temporary canal, the ductus venosus. All these extra passages close up soon after birth.

oxygen-depleted blood travels out in the twin umbilical arteries, which arise from the arteries supplying the legs and lower torso. Oxygen- and food-rich blood returns from the placenta in the umbilical vein, which veers upwards towards the liver. A special extension of the umbilical vein, the ductus venosus, traverses the liver and releases the life-giving blood into the vena cava returning to the heart.

Soon after birth, the umbilical vessels are cut or torn and they immediately close themselves off to stop the baby bleeding to death. The special bypass of the liver by the umbilical vein is now no longer necessary and the ductus venosus is gradually closed off. This also has the effect of forcing the blood returning from the intestines to go into the liver. This 'hepatic portal' circulation will be extremely important when the gut starts to absorb food. The hepatic portal system allows the liver to soak up any foods or poisons absorbed in the gut before they can damage more sensitive organs, such as the brain.

With the completion of her circulatory changes, our baby's adaptations to the outside world are almost complete. Her lungs are freely absorbing oxygen into her newly separated circulation and soon her gut will start to absorb nutrients from milk. Babies have yet one more trick to help them cope with the outside world, but unlike the changes in their lungs and circulation, they use this trick only if times get hard.

Newborn mammals are extremely vulnerable to cold. Not only have they just left a temperature-controlled incubator, but they are small, inactive and wet. To meet the challenge of staying warm in her new hostile environment, our baby has a unique tissue called brown fat. Brown fat is distributed over her shoulders and neck rather like a cape, and also around her kidneys. If she gets chilled, her nervous system responds by releasing adrenalin-like hormones that effectively 'switch on' brown fat. Brown fat is brown because it is packed with mitochondria – the cellular energy powerhouses that we mentioned in Chapter 1. Adrenalin drives these mito-chondria to start furiously burning metabolic fuel, churning out heat to warm up the baby. Despite its life-saving potential, brown fat is very much a temporary emergency measure and it is present only immediately after birth, perhaps as some sort of evolutionary recognition that this is a particularly chilly time. Whether it is used or not, the brown fat shrinks and disappears within a few days. From now on, the baby will have to shiver just like the rest of us.

Babies are not sexually quiescent creatures: they are planning ahead. Before and after birth, their reproductive organs are in a frenzy of activity. All of the baby's own eggs were generated while she was still in her mother's womb, and hormonally too she has been surprisingly active. During the nine months of her own gestation, she may have secreted more estrogen than she will produce during the whole of the rest of her life. The sudden cessation of hormone secretion at birth can have dramatic effects on her, and female babies can lactate or even menstruate in response to the sudden hormone withdrawal that overcomes them. After this time of hormonal flux, however, secretion of reproductive hormones gradually wanes over the first couple of years of the baby's life until she enters the reproductively quiescent phase of childhood. Only when her sexuality is reawakened at puberty will the elaborate machinery of procreation switch on once more.

Becoming Mum

Life does not change only for the baby after birth. Great change lies ahead for her mother as well. With birth, the mother's internal infant burden is replaced by an even more demanding external one. She now has to make yet another transition. Nine months ago, she changed from a sexual creature to a nurturing, pregnant one. Now once again, the mother must change direction, this time devoting her efforts to caring for her newly born infant. The mammalian mother is truly one of the triumphs of nature, switching seamlessly between the three roles of sex, pregnancy and childcare.

The first thing most mammalian mothers do once they have revived their babies is that they eat the placenta. It is not entirely clear why they do this, but it has been suggested that it allows them to garner valuable protein or useful hormones. Alternatively, they may just be disposing of a dangerous, predator-attracting liability. Almost all domesticated mammals, with the exception of mares, eat their placentas given half a chance. Do people naturally eat them too? Every so often, placenta eating drifts back into fashion, but it never really catches on in a big way.

There is no biological reason why people should not eat their own placentas – their digestive system seems better adapted for the job than that of, say, the afterbirth-devouring but otherwise vegetarian cow. Perhaps culinary aesthetics are more important for people than they are for other animals. Placentas simply do not look good to eat. Nor do I think that one's own placenta seems inherently more appetising. I had seen many placentas by the time I saw the one that came attached to my own daughter, but I cannot say that I immediately identified with that one as my own, and saw it as more succulent. My role in its creation did nothing to make it any more enticing.

Of course, every new mammalian mother has one job that is usually far more important than all her others. Provision of milk is the defining feature of mammals – *mammae* is Latin for 'breasts.' All female mammals, whether pouched, egg laying or whatever, can lactate to nourish their offspring. Millions of years ago, our distant ancestors began to feed their infants with the secretions oozing from glands in their skin. With the slow plod of evolutionary time, more and more of these glands were recruited for lactation and their outlets were gathered into a series of teats emptying out on the mother's belly. These teats were arranged in two neat lines, one on each side running from the mother's armpit to her groin. This is the layout of mammary glands that has persisted in all modern mammals. Some mammals have retained them all, such as sows, whereas others have kept only a few in the abdominal (cattle) or chest (humans) regions. The 5 per cent or so of people with extra nipples always have them somewhere along the two ancient nipple-lines.

Milk is the only sustenance of most newborn mammals, so it is a carefully concocted mixture of everything that a baby needs. Not surprisingly, these needs differ between different species and so there are as many different milks as there are types of mammal. Some babies grow quickly and some grow slowly; some live in dank, warm forest and others in dry, frozen wastes. In some mammals, such as whales and seals, milk is extremely high in fat to promote formation of blubber in the baby. By contrast, human milk is more at the low-fat, high-sugar end of the spectrum, and presumably this reflects our own particular needs. Milk contains not only nutrients in the form of protein, carbohydrate, fats, minerals and vitamins, but also a poorly understood admixture of hormones, antibodies, enzymes and cells. We do not know what many of these 'additives' actually do, but they are the components lacking in artificial milks.

The main advantage of breastfeeding is probably the short-term protection it gives babies against infection. As birth approaches, antibody-producing lymphocytes congregate in the breast, and after birth they secrete large amounts of antibody into the milk. The baby absorbs some of these antibody molecules, but some remain in the gut and inactivate many of the bacteria and viruses that cause neonatal diarrhea. Breast milk is now also thought to stimulate the baby's intestines to produce their own antibodies as well. These effects probably explain why breast-fed babies are less ill during their first six months than artificially fed babies.

Breastfeeding may also confer more long-term protection against other diseases, but the evidence for this is less clear. Breast-fed infants have been reported to be less likely to develop some allergic diseases by the time they become adults than bottle-fed infants. There is also evidence to suggest that breastfeeding may even reduce a baby's chances of dying from coronary heart disease when it grows up.

The results of a controversial British study published in 1992 suggested that breastfeeding may also cause children to develop a higher intelligence quotient (IQ) by the age of eight. The researchers found that this effect was still apparent even when they accounted for the parents' socioeconomic group (women in higher socioeconomic groups are more likely to breastfeed in most developed countries). To support this, they also pointed out that babies whose mothers tried to breastfeed, but failed, did not have higher IQs – so it is not an effect conferred by having the 'sort of mother' who wants to breastfeed. The effect did not seem to be due to the act of suckling itself, as children fed breast milk from a bottle showed the same increase in IQ as breast-fed infants. Not surprisingly, these results have proved to be very controversial and other groups have tried to repeat them. Many of these other groups claim that the link between breast milk and IQ is either very small or due to other factors.

Strangely, nutrition is not the only role for lactation, or sometimes even the most important. Suckling also helps to establish and maintain the psychological bond between mother and child. In fact, in species where babies are born in a very mature state and are capable of taking solid food immediately, bonding may be the main function of suckling. Another effect of lactation is to reduce the fertility of the mother, presumably to protect her and her infant from having to cope with the strain of another pregnancy too soon. Probably, the hormones produced to make the breast produce milk also suppress secretion of gonadotrophin hormones from the pituitary. Without gonadotrophins, the ovary cannot produce large ovarian follicles, so the menstrual cycle is jammed. Yet the contraceptive effect of lactation is variable between different species. In some animals that lactate for a relatively short time, lactation-induced infertility is fairly effective, but for mothers who produce milk for longer it can be more hit-and-miss. Also, as most female mammals only mate during their periods of 'heat', if lactation stops them coming on heat, then they are unlikely to get pregnant. This mechanism does not work for women – they can ovulate without realising it and become pregnant at any time during lactation. One of the most important messages that doctors can impart to new mothers is that they should not consider themselves miraculously infertile simply because they are breast-feeding.

The breast itself is an unusual organ: inactive for much of the time, its main constituent is fat. Also, it is unusual for such a delicate organ to be suspended unprotected on the outside of the body. The breast is designed to be an almost temporary structure and its vulnerable, unsupported position is made worse by a relative inability to repair the damage done by day-to-day wear and tear and the strains of intermittent lactation. Because of these design faults, and to the considerable financial benefit of the lingerie industry, the shape and integrity of the unsupported organ deteriorates progressively throughout a woman's adult life.

How is this strange, fragile tissue converted to a milk factory? It turns out that, as with so many aspects of reproduction, the process of lactation is orchestrated by hormones. Surprisingly early in pregnancy, the steroid hormones estrogen and progesterone help the mother to plan for her future baby. Under their control, a dramatic change starts to occur in the breasts. First of all, their blood supply increases, often causing obvious enlargement of the veins running just below the skin of the chest. Within the breasts, there is a gradual increase in the amount of glandular tissue, and this will eventually constitute much of the mass of the breasts when they are producing milk. From the main duct draining to the nipple, an extensive tree of branching tubes is constructed that permeates deep into the breast tissue and ends in thousands of tiny blind-ending milk-producing sacs.

Other hormones also encourage breast maturation, including prolactin from the pituitary (Figure 1) and placental lactogen from the placenta. These hormones help to adapt the different parts of the glandular tree to their particular roles: some cells secrete milk whereas others massage the milk along the tubules to the teat. In addition to all these changes, lymphocytes are actively drawn to the breast so that they can produce antibodies after birth to go into the milk.

Although much of what I have said about the maturation of the mammary glands applies equally to humans and animals, there is one unusual feature of human breasts. The human mother's preparations for producing milk are rather premature – the breasts are tuned up and ready to go by the fifth month of pregnancy. It is hard to see why human mothers are prepared for lactation so early because, without medical intervention, any baby delivered at this time would have absolutely no chance of survival. In many mammals, breast maturation is either a more leisurely or last-minute affair. Here we have yet another inconvenient quirk that

239

evolution has imposed on human mothers – the need to carry uselessly matured mammary glands around for most of pregnancy.

Although the breasts are able to secrete a small amount of milk before birth, the mother uses the end of pregnancy as a cue to boost production to large quantities within a few days. There is tremendous pressure on the mother for this process to work – the survival of her baby depends on it. There are two main changes around the time of birth that the mother uses to switch on her mammary glands. The first is the precipitous decline in circulating steroid hormones when the placenta is discarded, which acts as a strong drive to lactation. Although estrogen and progesterone cooperate in stimulating the development of the mammary glands during pregnancy, they also act on the breast to suppress the actual production of milk. Making milk is an expensive business, and the last thing a pregnant woman wants is to start making it when her baby has not even been born. The decline in steroid levels around birth, especially that of progesterone, also allows pituitary secretion of prolactin to increase, and it is this hormone that will be the prime mover in lactation from now on. Other hormones, such as pancreatic insulin and adrenal cortisol, have a role, but prolactin is the key to milk production.

The second change that helps lactation to get under way is an obvious one: the mother now has a hungry baby. The simple act of suckling stimulates nerves in the breast that are wired up to the brain to cause the release of more prolactin. When the baby removes milk from the breast as it feeds, the gland cells are also encouraged to refill the duct tree – a simple demand–led system. So the baby constantly strives to maintain its own food supply. This is why infants can be weaned off breast milk at any age. Any reduction in suckling soon leads to a waning, and eventually a cessation, of milk production.

Suckling itself is a remarkable process, a feeding mechanism unique to babies. Human babies are not thought actually to suck milk from the breast, at least not if their mother wishes to avoid considerable pain. Instead, the nipple is drawn deep into the mouth so that it is lodged far back in the throat. The baby's tongue then squeezes the milk out of the portion of the breast held in its mouth by massaging it against the roof of the mouth. This system is very efficient and does not seem to require the application of suction to the delicate tissues of the nipple.

Suckling has effects far beyond the simple withdrawal of milk. Not only does it encourage the production of prolactin by the front part of the pituitary gland, but also it induces secretion of oxytocin by the back of the pituitary. Oxytocin, not satisfied with its roles at the start and end of pregnancy, is also an important hormone for driving lactation. Oxytocin makes the small milk ducts of the breast push milk into the portion within the baby's mouth – a process called milk ejection. In a way, a suckling infant is ordering its current meal with oxytocin and booking its next one with prolactin.

Another feature of the milk ejection reflex is that it can be 'conditioned'. Conditioned responses are ones that can be retrained to react to cues other than the one for which they were originally designed. Because of this, many lactating women eject milk at the mere sight or sound of a baby, just like Pavlov's dogs drooling when they heard the dinner bell. The ejection response has another job, as oxytocin also helps the newly empty uterus to shrink after birth. Many women feel these contractions and may even experience uterine pain when they first feed their newborn babies, but they can take heart from the fact that the whole experience is probably helping them to regain their slimmer pre-baby figure.

The oxytocin released during suckling has one more effect: it strengthens the bond between mother and baby by acting directly on the mother's own brain. There seem to be few limits to the diverse uses to which the human body has put this hormone. The maternal bonding action of oxytocin may even have been extended to adulthood, and some scientists are now studying the role of this hormone in sexual attraction and pair-bonding. Oxytocin has been called, with some justification, the 'hormone of love'.

Lactation continues for differing periods of time in different mammalian species, but in humans it often ends, or rather is stopped, some time towards the end of the first year of life. The weaning process can be either abrupt or gradual, but either way, the breasts can sense that less milk is being extracted and they scale down their production accordingly.

Lactation can place tremendous strain on a mother – deer hinds expend more energy during lactation than during pregnancy, and this is often the case in other species too. Milk production seems to be a robust process, continuing even if the mother is thin or malnourished, although even a well-fed mother's metabolism can scarcely keep up with the demands placed on it by a hungry baby. A particular drain is the huge amount of calcium and phosphorus put into the milk to help the baby's bones grow. Although the mother's kidneys adapt to retain more of these substances, she has to dip into her own reserves of minerals all the same. As a result, her bones suffer mineral loss by the time of weaning.

Given that lactation is such an exquisitely controlled mechanism, it is difficult to explain why humans find suckling so difficult. Most mammalian mothers and babies seem to know exactly what to do – the baby soon learns where to go for a meal and the mother is instinctively able to cooperate. In comparison, women often seem to have a great deal of trouble starting to breastfeed. Although newborn babies know how to suckle, they frequently do not seem to know how to get attached in the first place. Similarly, women often do not instinctively know how to help their baby to latch on to the nipple. Many are surprised at just how much of the breast has to be stuffed into their infant's mouth for pain-free feeding to be possible. Painless suckling can take place only when the baby feeds without using suction, but improperly attached babies still try to feed by sucking. If sucking leads to pain for the mother and often the eventual end of breastfeeding, why do babies attempt it? Suckling does not seem to be at all painful in other mammals, even for the most nervous and inexperienced mothers, at least until their babies' teeth start to grow. In countries where breastfeeding is the only sanitary way for newborn babies to be fed, the problematic nature of human lactation is a matter of life and death.

As if the pressures of having a new baby were not enough, most mothers also have to cope with some form of psychiatric problem thrust upon them by birth. These 'postnatal mood disorders' are a

challenge to psychiatrists because their causes are not fully understood, even though they are among the most common of all
psychiatric problems. Also, they vary so much in their severity in
different women – from almost imperceptible to life-threatening
to both mother and baby – that it is sometimes hard to believe that
these disorders are related to each other at all.

The mildest form is usually called postpartum blues, and it
probably affects 70–80 per cent of new mothers. Its characteristic
feature is that it starts extremely quickly and it does not last long –
perhaps from the second to the fifth day after birth. Women with
postpartum blues suffer from a disorientating mixture of mood
swings, poor concentration and anxiety. The causes that are usually
suggested are sleep deprivation, the shock of suddenly having to
become a parent, the withdrawal of placental hormones and a
readjustment of pituitary hormone secretion. When the causes are
spelt out like this, it hardly seems that postpartum blues is a
'disorder' at all – I suspect that most people who were in such a
state of flux would be a bit tetchy and confused. Perhaps
'postpartum complete exhaustion' would be a better term.

Postnatal depression affects around 10 per cent of mothers. The
symptoms of postnatal depression are like a much worse form of
postpartum blues, but with the additional features of insomnia and
loss of libido. It can also start much later – up to six months after
birth. The causes of postnatal depression are controversial, and this
has made it difficult to decide how to treat the disease. The
disorder has often been claimed to be caused by the dramatic
decline in circulating steroid hormones after birth, and this idea is
supported by the fact that estrogen treatment relieves its
symptoms. This does not, however, prove that estrogen 'deficiency' causes postnatal depression – penicillin may cure infections, but infections are not caused by lack of penicillin. One factor
that has been consistently linked with postnatal depression is lack of
emotional and practical support for new mothers. Also, doctors
sometimes remark that the disorder often occurs in previously
confident, assertive women, and it has been suggested that these
women are less able to accept the fact that babies often behave in
frustratingly illogical ways. Some psychiatrists suggest that fathers

can more easily accept that a screaming baby is not a reflection of their own competence as parents, and that this can seem to mothers like a lack of commitment to solving the 'problem'.

First described by Hippocrates, the most severe postnatal mood disorder is postnatal psychosis. Unlike healthy people, or people with neuroses, people suffering with psychoses appear to have a distorted view of reality, often leading to delusions and extremely erratic behaviour. Postnatal psychosis usually develops within the first three weeks after birth and affects between 0.1 per cent and 0.2 per cent of mothers. It is an extremely dangerous disorder and can lead to suicide, child neglect, child abuse or even infanticide. The first stage of treatment is to protect both mother and baby. The condition can be treated with antipsychotic drugs, but these can enter the milk and so breastfeeding must be stopped. Fortunately, around three-quarters of sufferers recover, but a considerable number relapse if they give birth again. Relapses are very rare if pregnancy is avoided. It is difficult to explain how something like postnatal psychosis could be caused by birth, but it is a very real disease, and the hunt is on to find the factors that trigger it.

Quite apart from the demanding job of producing and feeding a baby, it seems that the responsibilities of motherhood never really end. Recent research into evolution of animals' life histories has suggested that childbearing affects mothers for the whole of their lives. These effects of pregnancy are built into the entire biological fabric of being a woman, and happen whether or not she has children.

Unusually among mammals, women undergo a definite menopause. Whereas many female mammals' fertility gradually wanes as they age, women's fertility ends relatively abruptly in their forties or fifties. Menopause happens because a woman's ovaries simply seem to switch off. They stop ovulating and stop producing steroid hormones, and as a result, menstruation ceases as well. The pituitary does not seem to be 'expecting' the menopause and it responds by pouring out huge quantities of gonadotrophin hormones in a desperate attempt to restart ovarian activity. Within a few years, women are transformed from being fully fertile to being completely infertile.

There is considerable debate over why women discard their fertility in this off-hand way, largely because it is one of the most unusual and dramatic features of the human life plan. Like the questions of why women menstruate and why eggs are bigger than sperm, we have no definitive answer, but the mystery of the menopause may be partially solved by looking at how evolution makes animals the way they are.

Some of the suggested causes of the evolution of the human menopause are rather negative. According to these ideas, menopause is not actually *for* anything. One idea is that women were never designed to live as long as they do now, but were instead meant to collapse, exhausted, before the menopause ever had a chance to occur. By this argument, the menopause did not evolve by natural selection, because women were usually dead long before menopause could occur. So, menopause has no function – it is just what happens when you let women live beyond the life span for which they were designed. There are, however, a couple of problems with this idea. First, if women were under no pressure to evolve a cessation of their fertility, then why have they ended up with exactly that – an unusually dramatic and abrupt menopause? Secondly, anthropologists now believe that pre-agricultural humans (which is what we were until relatively recently) often lived well beyond their forties. Surprisingly, it is thought likely that human life expectancy probably fell when people first came together into agricultural communities. All this pre-agricultural longevity makes the women-never-lived-beyond-thirty argument seem less appealing.

Another 'negative' suggestion for why the menopause evolved is more promising. The argument runs like this: human mothers put an especially large investment into each of their children, mainly because they are extremely vulnerable and they grow up very slowly. Because of this spectacular investment, hunter-gatherer women probably could not produce a large number of children. In fact, most of their children who were destined to survive would be produced early in life, meaning a woman's total number of children who reached adulthood was largely decided by the time she reached thirty. So what happened to a woman after thirty was irrelevant as far as future generations were concerned, and by random chance women acquired a process of uncontrolled reproductive deterioration. This came with all sorts of unpleasant side-effects that could not be weeded out by evolution, because natural selection did not act on women who had already produced all their offspring. While this theory seems rather unfair, misogynistic even, it has some advantages. Portions of animals' life histories that escape natural selection do indeed seem to accumulate useless 'junk' features. There are still lingering doubts about this theory, however. Perhaps menopause is too active and definite a process to have evolved by chance, without natural selection.

There are more 'positive' ideas about why menopause evolved, and they can often seem more convincing, although of course this does not mean they are correct. One is that menopause may have evolved to reduce the incidence of congenital diseases more common in babies born to older mothers. We saw in Chapter 3 that females establish a bank of eggs in their ovaries before their birth and that this bank is the source of all the eggs they will ovulate in their adult life. By the time they are ovulated, these eggs will be decades old, and they can accumulate damage as they age. For example, this 'stale egg' process is probably the reason why Down's syndrome is more common in children born to older mothers (the incidence is about 0.04 per cent in mothers under twenty and 2 per cent in mothers over 45).

One thing that the idea has in its favour is that it should work – presumably menopause does reduce the incidence of Down's and similar syndromes. On the other hand, there are also reasons why menopause might not have evolved to stop women using 'stale eggs'. Women who have children when they are forty do not consistently produce children with genetic diseases. In fact they rarely do, even though their chances are higher than their younger counterparts. A woman who produces a baby later in life still has a pretty good chance of having a perfectly healthy one. From the point of view of a woman's reproductive success (the number of babies she rears to adulthood), it seems unlikely that she would give up that chance because of the slight risk of a genetic disease. After all, babies get genetic diseases for reasons other than the age of their mother, and there is no special mechanism for weeding these out of the population – they just take their chances with natural selection like everyone else. Surely there has to be a better reason for the evolution of something as distinctive as the menopause?

There is one more theory that has been put forward to explain the menopause, and like the last one it proposes a real, beneficial purpose. Researchers have tried to explain the menopause in terms of how it might increase the number of children women rear to adulthood and how natural selection can act on a woman after she has delivered her last child. In an attempt to reconcile these two questions, it has been suggested that a post-menopausal woman has a definite evolutionary role, which is made easier by her new-found infertility. The obvious contribution that a post-menopausal woman could make to the number of genes she passes on is caring for the children she has already produced.

A human mother's contribution to her children's future is far from over when they are weaned. In fact, most of the effort she puts into rearing them to maturity is expended in the fifteen or so years after weaning. Perhaps the slow development of human children has meant that there is pressure on women to stop producing new babies halfway through their life so that they can better care for the babies they have already produced. Natural selection may have inflicted the menopause on women to stop them placing the success of their existing children in jeopardy. Evolution forced women to 'quit while they are ahead' and prevented them from producing an unmanageable number of children.

Whatever the cause of the menopause, the further we depart from our natural hunter-gatherer lifestyle, the more it becomes an inconvenience rather than a boon. Socially, as career pressures encourage women to reproduce later in life, the whole reason for the menopause has ceased to exist, and instead it has become a feared barrier to fertility. Women who want children must balance their career aspirations against the timing of the impending onset of infertility, which of course cannot be predicted with any confidence.

If menopause evolved to stop women compromising the

children they produce in early adulthood, then pregnancy can be seen as a process of overwhelming importance in a woman's life. The ability to produce children has been woven deeply into the fabric of what it means to be a woman. Even after the menopause, a woman cannot escape the burden that nature has placed on her. Although she is no longer fertile, her whole biology is still designed to help her to rear any children she may have had. It is almost as if the spectre of pregnancy haunts women throughout their lives, long after they are no longer of 'reproductive' age, and even if they never conceive a child.

The ability to produce children affects women for the rest of their lives. Our story of pregnancy, which started with the miraculous process of conception, has taken us as far as the death of the child's mother. All that now remains is the mother's genes, trickling merrily through the generations.

Babies in bottles

I predict that in one hundred years' time, human pregnancy will be very different from the way it is now. To the likely surprise of many, pregnancy is set to become a controversial subject, in much the same way that fertility was in the last thirty years of the twentieth century.

Medicine is, by definition, a profoundly unnatural process. Within the animal kingdom, humans are unique in trying to cure disease and care for the afflicted. By eliminating and alleviating most of our diseases, we have effectively freed ourselves from the pressures of natural selection. I am not saying that people do not inhabit a competitive world, but we now experience very different stresses from those of humans a thousand generations ago. Medicine has largely removed disease from the position it used to occupy as the great arbiter of life and death in human communities. Most people are now partially freed from the ravages of uncontrolled disease.

Medicine has also allowed us to escape the shackles of natural selection in another crucial way – by giving us control over our own fertility. Again and again, medical technology has allowed us (well, women mainly) to escape from the forces that once controlled our lives. I like to think of these as the four tyrannies from which we are constantly striving to extricate ourselves: the tyrannies of lactation, childbirth, conception and pregnancy.

The first tyranny we fought was probably the least oppressive: lactation. The necessity to breastfeed our children may be an inconvenience rather than a scourge, but women throughout history have still done their best to avoid it – there was a word for 'wet nurse' in the Sumerian language used 4000 years ago. Whether it be wet nurses or artificial milks, human ingenuity has given us ways to do the unnatural thing, and avoid breastfeeding. As this is the least of the tyrannies, it is perhaps little surprise that artificial milk has been the first reproductive artifice to suffer a reversal in medical thought. Even in countries where formula milk can be made up cleanly, medical authorities are now pressing for a return to breastfeeding.

Childbirth was probably the next reproductive tyranny against which humans rebelled. Until 200 years ago, many women and babies died during childbirth, and birth was perhaps the most dangerous time in any mother's life. As we have seen, advances in obstetrics and the development of the Caesarean section have removed many of the risks associated with childbirth. Modern obstetric practice is, of course, profoundly unnatural, but it has brought welcome relief to millions.

To date, our assault on the third tyranny, that of conception, has been our most impressive and perhaps our most important too. In the twentieth century, fertility changed from an uncontrollable force of nature to a facility that couples can control almost at will. The main way in which we can control conception is by suppressing it, and new methods of contraception must rank as the greatest inventions of the twentieth century. In the past fifty years, barrier methods have been supplemented with hormonal contraceptives of amazing efficiency. 'Family planning' is the perfect term for our invention, as it shows how far we have changed the natural state of affairs. Natural selection plans most animals' families for them. We can plan our own.

Of course, humans want to have their cake and eat it. Not only do we want to suppress our fertility when it is not wanted, but we also want to be able to improve it when it is poor. The field of reproductive technology took tremendous strides in the twentieth century, as the techniques of artificial insemination and embryo transfer, first developed in animals, were applied to people. An entire *in vitro* fertilisation (IVF) industry arose, dedicated to overcoming infertile couples' problems. Women with blocked Fallopian tubes can now often become pregnant if the obstruction is bypassed by *in vitro* fertilisation of eggs with ejaculated sperm outside the body.

As reproductive technologists have become more skilful, they can now treat forms of infertility once thought completely intractable. The main recent beneficiaries have been infertile men, as techniques have been developed that allow direct injection of sperm into the egg or under the zona pellucida (intracytoplasmic sperm injection and subzonal sperm injection; ICSI and SUZI). Doctors are now hunting ever further for the last bastions of infertility to heroically attack. For example, research is now under way to try to persuade non-germ cells to change into sperm and eggs under artificial conditions. This possibility would give hope to the most problematic cases of all: people rendered infertile by the absence or loss of ovaries or testicles.

The progress of reproductive technology has been truly remark-able in its rapidity and scope. Further improvements are no doubt on the way, and perhaps one day the third tyranny of conception will have been entirely vanquished. How much more can we free ourselves from the reproductive responsibilities that nature has placed upon us? When we can conceive at will, avoid the experience of birth and feed our babies artificially, how else will we want to avoid reproducing naturally? I think it is inevitable that eventually, and probably quite soon, people will want to avoid the burden of pregnancy itself. As the fourth reproductive tyranny, pregnancy will one day be considered as unnecessary as danger-ously protracted labour is considered today. I doubt that ethical considerations will restrain us for long – there was tremendous ethical opposition to IVF in the 1970s and 1980s, but look how all-conquering the procedure has now become.

The idea that women can avoid pregnancy altogether may seem like an idea too far. However, the huge medical-scientific-research behemoth is already making gradual, inexorable moves in precisely that direction. Paediatricians are now able to save the lives of premature babies born at ever-earlier stages of pregnancy, and, in some hospitals, babies can be fairly reliably rescued when they are born at twenty-eight weeks – something almost unthinkable a few decades ago. At the other end of pregnancy, scientists are developing the skills they need to grow early embryos under artificial conditions for longer and longer (although in many countries it is forbidden by law to attempt this in humans). The gradual 'erosion' of the two ends of pregnancy raises an obvious question: what happens when fertility experts can grow human embryos to the stage when paediatricians can transfer them safely into the outside world? Will pregnancy no longer be necessary?

Many people have suggested this possibility before. As early as 1923, the evolutionary biologist and celebrated Marxist J.B.S. Haldane even gave extra-uterine generation of babies a name: 'ectogenesis'. Many people were introduced to the idea of ectogenesis by a memorable passage in Aldous Huxley's *Brave New World* in which all humans were grown in bottles on special baby farms. Ectogenesis, while hardly a mainstream research area, has been at the backs of the minds of many biologists over the past hundred years. I mention it here because I believe I am writing at the time when the pioneering age of ectogenesis is just beginning. Almost every other aspect of human reproduction is now under, or will soon be under, our control. The ability to produce babies without pregnancy is the obvious next step. I am not saying it is right, I am simply saying it is obvious. And, I would suggest, inevitable.

Reports have appeared in the scientific literature that sound very like attempts at ectogenesis. One of the first attempts at maintaining the human fetus under artificial conditions outside the womb was by Robert Goodlin at Stanford University in the mid-1960s. His work was extremely controversial as it involved the maintenance of human fetuses miscarried at fairly advanced stages of pregnancy. Although intended as a research programme to help premature babies, Goodlin's work was a public-relations disaster and he was forced to end the study in the face of considerable public hostility.

After Goodlin's experiences, scientists interested in maintaining fetuses out of the womb concentrated on studies in animals. Although their intentions are usually to advance our understanding of premature babies' needs, their work is gradually edging us nearer to complete ectogenesis. For example, in 1984 a research group at the University of Western Australia demonstrated that they could grow twelve-day-old rat embryos for twenty-four hours by incubating them in artificial fluid (rat pregnancy lasts three weeks).

Most surprising of all are the studies carried out in the 1990s by Japanese groups on maintaining goat fetuses in artificial surroundings. The work of Yoshinori Kuwabara and his colleagues at the Department of Obstetrics and Gynaecology of the University of Tokyo is especially novel because it has tackled the main problem that must be overcome to allow ectogenesis – the formation of an artificial placenta. To achieve this, these groups have devised a method of transplanting the fetal goat's umbilical arteries and vein on to a membrane suspended in fluid loaded with everything the fetus needs. The fetus and its new placenta are then kept at a constant nanny goat's body temperature in artificial amniotic fluid. Gases and nutrients can then be passed across this membrane sufficiently quickly to maintain the developing kid, sometimes for up to three weeks. These fetuses have much more room than they would have in a uterus, and to stop them overexerting themselves or swallowing too much fluid, they are sedated and partially paralysed. Remarkably, there is evidence that the fetuses' lungs grow and start to secrete surfactant during this period. The kids can be 'born' and survive for some months outside their incubator, but they suffer from the long-term effects of the drugs used to restrain them.

Whatever one may think of the ethics of carrying out these experiments, it seems clear that ectogenesis is more a question of 'when' than 'if'. How will we cope with the ability to generate children outside our own bodies? How will it affect the way we see ourselves? Of course, there will be great public opposition to ectogenesis, but somewhere, someone will want to do it. How will we react when it has been achieved for the first time? There was great opposition to IVF when it was first carried out successfully, but this opposition gradually faded away until IVF now seems the obvious choice for infertile couples.

People will probably oppose ectogenesis while they can see no need for it, but the situation may change when it eventually becomes safer than making babies in the womb. And what of women with no uterus − will we still condemn them to eternal childlessness? I suspect that the time will come when people will accept ectogenesis, regardless of how 'Frankensteinian' it may seem now. Someone will want it badly enough. History has taught us that ethical objections to new procedures soon wither away as these procedures lose their novelty. All too often, ethical objections have simply turned out to be fear of the unknown, or just plain chauvinism against novelty. So-called ethics can easily become a disguise for simple intellectual inertia.

Medicine has left us in a strange position. I said at the beginning of this book that pregnancy was the most 'visceral' part of our modern life. Despite this, we humans are continually striving to de-visceralise even the production of our own children. This process is not right or wrong; it is simply what we are doing. Soon we may be left with controllable conception, no childbirth, no breastfeeding and no pregnancy. Perhaps people will look back to books such as this as charming historical oddities from a quaint era. As I finish this book, I wonder if perhaps we have come to the end of pregnancy in more ways than one.

But one seed of our ancestors will always remain within us. No matter how neat and tidy and reliable we make pregnancy, there is one urge we will probably never lose. This is the urge that controls the pattern of our lives, the urge that, paradoxically, will drive medicine to make pregnancy itself obsolete: we will probably always want to have children.

Notes

○ ○ ○ ○ ○ ○ ○ ○ ○

Chapter 1: Origins

p. 24: Mendel's peas

Here is a point-by-point summary of Mendel's results, and how they helped him to expound his theory of genetics. First of all, Mendel knew that pea plants come in tall and short varieties and that he could control their mating by either cross-pollinating or self-pollinating them. He found that:

1 Short plants pollinated by other short plants (or by themselves) always produce short offspring.

The converse was not true of tall plants:

2 Tall × tall crosses might yield all tall offspring or a mixture of tall and short.

However, some tall plants did reliably produce only tall offspring when crossed with themselves and each other, as if there was no 'shortness' within them. The progeny of these crosses would in turn produce only tall offspring. These findings suggested to Mendel that short plants were always 'pure' short, whereas tall plants may or may not be 'pure'. He solved the problem by fitting an elegant model to the outcomes of tall × short crosses:

3 A 'tainted' tall pea plant crossed with a short plant produced offspring, 50 per cent of which were short and 50 per cent of which were tall.

4 A 'pure' tall plant × short plant cross generated only tall progeny, although members of this next generation when crossed with themselves or each other *always* produced 75 per cent tall plants and 25 per cent short.

You may wish to pause at this point, look at the four statements above and try to work out for yourself the way in which height is inherited in peas, just as Mendel did.

Mendel's rules of inheritance are simple – each individual has two sets of instructions for any trait, one inherited from each parent. This individual will pass on one of these copies at random to its progeny. These instructions are called genes. Often one of a pair of genes can dominate the other, rendering it undetectable, or alternatively the effects of the two genes may 'mix'. That is all there is to it.

In the case of pea plants, the dominant gene is the one for tallness; let us call it *T*.

1 Any plant with at least one *T* gene will be tall.

2 Only plants with two 'non-tall' genes, *t*, will be short.

This *t* gene, which is easily dominated by the other is called recessive. Now we can work the whole problem out.

3 The short plants are *tt*, the pure-breeding tall plants are *TT* and the non-pure-breeding tall plants are *Tt*.

4 Short × short progeny can only receive *t* and *t* from their parents, and so must be *tt*: in other words, short.

5 Pure-tall × pure-tall offspring can only be *TT*, or tall.

6 Because plants pass a gene on at random to each offspring, non-pure-tall × non-pure-tall crosses (*Tt* × *Tt*) can be tall (*TT* or *Tt*) or, less often, short (*tt*).

p. 25: Sickle–cell anaemia

In sickle-cell disease sufferers, blood cells contain defective pigment and, because of this, they contort into a characteristic sickle blade shape when they squeeze into blood capillaries. Not only can the blood cells get jammed in blood vessels and cause oxygen starvation of body organs, but the cells are also liable to get damaged, leading to loss of cells from the blood (anaemia).

When an *Ss* male carrier and an *Ss* female carrier produce a child, they each give it one haemoglobin gene at random. There is a 25 per cent chance that they will both give it an *S* and it will be a healthy *SS* baby. There is a 25 per cent chance that the father will give an *S* and the mother will give an *s*, and a further 25 per cent chance that these contributions will be reversed. Thus the baby has a 50 per cent chance of being a healthy *Ss* carrier. Unfortunately, there is also a 25 per cent chance that it will inherit an *s* gene from each parent, and if it does so, it will be *ss* – a baby with sickle-cell disease.

Why has the defective *s* gene persisted in the human population when a proportion of *s*-bearing individuals die before reaching the age at which they could produce their own children? Surely over time the *s* gene should get filtered out of the population? At least it should not have built up to its present abundance? The answer is that, rather perversely, the *s* sickle-cell gene confers an advantage on the people who carry it. It is suspected that *Ss* individuals are more resilient to malaria than *SS* people. Suddenly it all makes sense. The gene for a dreadful inherited disease is maintained in the human population because it gives improved resistance to one of the major killers of mankind. Nature seems to be extremely inventive.

p. 28: Why you can never pass on all your genes

You can never pass on all your genes to your children, no matter how many you have, and this is why. If you have one child, it will inherit approximately half of your genes. If you have a second child, then this too will get half of your genes. However, because you pass on these genes at random, around half of the genes that the second child gets will have already been given to the first child too. Hence, by the time you have your second child, you will have passed on roughly three-quarters of your genes. If you have a third child, then three-quarters of the half of your genes that you donate to it will already have made their way into your first and/or second children. Because of this, you have now passed on seven-eighths of your genes. After your fourth child, you will have passed on fifteen-sixteenths, and so on.

It is a law of diminishing returns: the more children you have, the more of your genes are passed on, but you can never transmit them all, no matter how many you have.

p. 29: Ploidy
Having two copies of each gene is called diploidy – we are diploid. Having one copy of each gene is called haploid – sperm and eggs are haploid because they contain only half the genes of the individual that made them. When a sperm and egg fuse, they reconstitute a new diploid individual.

Other creatures, more adventurous than ourselves, occasionally experiment with triploidy, tetraploidy and so on.

p. 32: Vertebrates
To be more precise, the vertebrates are:

1 Fish: bony fish, sharks and rays.
2 Amphibians: frogs, salamanders and some other, stranger things.
3 Reptiles: an assortment of tortoises, lizards, snakes and crocodiles.
4 Birds: feathers, beaks, wings.
5 Mammals: animals that suckle their young, including us; usually furry.

p. 34: Rock lizards
If you have ever bemoaned how global culture is reducing mankind to a homogeneous soup, just be glad that you are not a Caucasian rock lizard. These little creatures are parthenogenetic, but the whole species appears to be a single clone. Thus, they are probably the only vertebrate species that consists entirely of identical individuals.

p. 42: Small eggs do not necessarily mean small offspring
The principle that larger eggs are able to generate larger embryos does not apply to all animals, however. Certainly, in animals that do not undergo prolonged pregnancy – birds, for example – the hatchling cannot be any larger than the egg that produced it, and so it is advantageous for females to produce as large an egg as possible. In animals that do exhibit pregnancy – mammals, for example – this is not the case. Our embryos are nourished by their mothers for a prolonged period before they are born, and so they grow to be far bigger than the egg that gave rise to them. Hence, there is far less pressure for mammalian eggs to be large. This is why hen's eggs are around 80 million times heavier than a human egg.

p. 44: Do we inherit only maternal mitochondria?

Recent evidence suggests that we do occasionally inherit mitochondria from our fathers, but this is an extremely rare occurrence.

p. 48: Why the male fly's bicoid gene is functional

Although the product of the *bicoid* gene is made by female flies and injected into the egg to exert its effects, this does not mean that male flies' copies of the gene are irrelevant. Although *bicoid* genes are inactive in the male fly, they are passed on to his offspring in the usual way, and half of these offspring will be female. These offspring will then use the copies of the *bicoid* gene they inherited from their fathers. Thus, a male's *bicoid* genes are used to make his grandchildren.

There are many examples in nature of genes that are present in both sexes, but which are only used in one. For example, men carry all the genes for making a uterus, they simply do not use them. These genes may be passed on to, and be used by, any of their female descendants.

Chapter 2: Breaking the cycle

p. 70: Estrogen in the news

Estrogen has been in the news recently in the context of estrogen-like pollutants that are claimed to induce sex-changes in wild animals and estrogenic plastics claimed to cause reduced sperm counts in men. Estrogen has acquired this infamy because many artificial chemicals share features with estrogen and can exert estrogen-like effects on animals. Males in particular are often exquisitely sensitive to estrogen: a little estrogen goes a long way.

Used for some time as a major component of the contraceptive pill, estrogen has become even more medically important recently for two main reasons. The first is that it makes up a large part of hormone replacement therapy for post-menopausal women, especially because low doses of estrogen are effective in preventing osteoporosis, or brittle bone disease. Another medical innovation related to estrogen was the discovery of its role in stimulating the growth of breast tumours. Just as estrogen coordinates growth of breast tissue during pregnancy, it also drives the disordered proliferation of some mammary cancer cells. By chance, yew trees produce a substance called tamoxifen, which suppresses the effects of estrogen by stopping it binding to breast cells. By treating some women with breast cancer with tamoxifen, it is possible to remove one of the main forces driving tumour growth – their own estrogen.

p. 75: Progesterone in the news

Like estrogen, progesterone looks set for its share of controversy in the future. Because of its importance during the cycle and pregnancy, the development of a molecule called RU486, which blocks the effects of progesterone, could lead to drugs with many medical uses. first of all it seems that, as with estrogen, some tumours grow partly because of stimulation by progesterone. Because of this, the outlook for people with these tumours may be improved by treatment with

RU486. Secondly, progesterone blockers also have a great deal of potential as contraceptives. Not only could they be used to suppress the normal cycle, but also they could prevent implantation of newly fertilised embryos, rather like the present 'morning-after pill', but perhaps with fewer side-effects. The feature of these drugs that has caused the most argument is their likely ability to be able to induce abortion throughout pregnancy, and this has delayed the drug becoming licensed for other uses.

p. 76: The reproductive cycle of women and female deer, step by step

1 Follicles are chosen during a period of gonadotrophin/estrogen interaction.

2 Follicles enlarge under the control of gonadotrophins, although estrogen keeps gonadotrophin release in check.

3 Follicle estrogen secretion rises to a level where it starts to stimulate gonadotrophin release.

4 Blood estrogen and gonadotrophin levels both 'surge' and ovulation occurs.

5 The old follicle collapses and forms the corpus luteum, which makes progesterone under the steady control of gonadotrophins.

6 At a pre-arranged time, the corpus luteum is destroyed (luteolysis) and blood progesterone falls.

7 Pituitary gonadotrophin secretion is relieved from suppression by progesterone – go back to step 1.

p. 86: Why marsupial reproduction is not inferior

Unfortunately, the unique position of marsupials within the mammalian scheme of reproduction has been overshadowed by arguments about the supposed inferiority of marsupial reproduction compared with that of 'placental' mammals. Long assumed to be an irrelevant evolutionary relic cast adrift on their island continent, and because they are immature, tiny and vulnerable at birth, our pouched cousins have often been assumed to have an inferior system of reproduction. Surely rearing your young in a pouch is a desperate attempt to make up for an inability to produce a decent placenta? However, many of the arguments for marsupial reproductive inferiority no longer appear to stand up to close scrutiny. For a start, some marsupials can and do develop perfectly functional placentas and some species undergo protracted pregnancies as well.

It has been suggested that female marsupials are somehow reproductively inadequate because they are anatomically unable to fuse the two sides of their uterus into a single large central canal, as in humans (kangaroos have two cervices!). While this fusion may indeed be impossible for them, very few 'placental' mammals actually show much fusion of the two sides of the uterus either. The human single-bodied uterus is the exception rather than the rule, and most 'placental' mammals actually nurture their young in two unfused uterine horns, just like marsupials. In fact, certain marsupials which obviously felt the same need as humans have rather ingeniously added an additional large central uterine canal over the course of their evolution. There seem to be few limits to the reproductive ingenuity of marsupials.

One fact that cannot be avoided, however, is that lactation is an inherently less efficient system than placentation – processing nutrients into milk and getting them absorbed by an infant's gut is simply more energy-intensive than passing them across a placenta. Marsupial females are probably willing to meet this additional energetic cost because of the flexibility that lactation gives them. Not to put too fine a point on it, if you are a female kangaroo and times get hard, you can just dump your pouch joey and wait for things to improve before you have another one. 'Placental' mammals cannot do this – once pregnant, women, bitches and hinds are committed to a prolonged period of intensive investment in their fetuses. The only option for them is miscarriage, which often occurs only when the mother is already *in extremis*. Thus, marsupial females have escaped the tyranny of the unstoppable pregnancy by perfecting lactation (which is, after all, the defining feature of mammals). Many marsupial females regulate their fertility by keeping or discarding pouch-young as appropriate. Women only achieved control over their reproductive destiny in the last few decades: kangaroos have been doing it for millions of years.

p. 86: How marsupials control the length of pregnancy

This is by no means the limit of marsupial ingenuity. Having exploited lactation far more effectively than 'placental' mammals, they have also exploited another handy aspect of pregnancy, known as embryonic diapause. Diapause is the ability of a mother to arrest the development of her embryo and store it in suspended animation before it has a chance to implant into the uterus. Many marsupial and 'placental' mammals undergo diapause because it allows them to disconnect the timing of birth from that of conception. Why should birth automatically follow conception by a fixed interval when this might conflict with the need for the young to be born in a particular season, or after an older sibling has been weaned?

Marsupials have developed an array of mechanisms to arrest embryonic development, but a typical one is to reduce progesterone secretion from the corpus luteum and to pump out prolactin from the pituitary. For example, soon after a wallaby conceives, it can arrest the growth of its new embryo, depending on the season or whether it is currently producing milk for another baby. The embryo is put into storage until a pulse of progesterone reactivates it. Because of this, progesterone is often released in two pulses during marsupial pregnancy: the first when the corpus luteum is first formed and the second when diapause is ended. Diapause is at the heart of reproduction in many marsupials. They are not just callous baby-dumpers when the going gets tough – they can also judiciously save up their embryos for better times.

p. 92: Roe deer can store their embryos, just like marsupials

There is one species of deer that undergoes embryonic diapause, just like the marsupials I mentioned earlier. Roe deer are Britain's smaller native deer and they are shy and solitary. In contrast to other deer, female roe deer ovulate only once a year and, because of this, roe deer have developed an elaborate courtship to ensure a male is present when this valuable ovulation occurs.

One crucial element of the luteal self-destruct system (uterine receptors for oxytocin) is missing in roe deer and so the whole multiple-cycle, maternal recognition of pregnancy system does not occur. So, whether the female roe is mated or not, she assumes she has an embryo in her uterus and does not cycle again.

It is remarkable that by suppressing the sensitivity of just one organ to just one hormone, the roe deer has converted its entire breeding system. It is no longer a red deer-like 'corpus luteum protector' but has become a kind of cross between a marsupial and a dog, a true 'assume you're pregnant' mammal. In fact, roe deer suspend their embryos like some marsupials, maintain luteal progesterone secretion throughout pregnancy like dogs, and do not have maternal recognition of pregnancy, like marsupials and dogs.

Chapter 3: Making babies

p. 110: Von Baer's ideas about evolution and development

' . . . the embryos of mammalia, of birds, lizards, and snakes, probably also of chelonia [tortoises and turtles], are in their earliest states exceedingly like one another, both as a whole and in the mode of development of their parts; so much so, in fact, that we can often distinguish the embryos only by their size. In my possession are two little embryos in spirit, whose names I have omitted to attach, and at present I am quite unable to say to what class they belong.'

p. 129: The inverted insect

The human embryo has a head end and a tail end, a gut tube running along its belly and a nerve tube running along its back. This is the opposite arrangement from what we find in insects: they have the nerve tube running along their belly and a gut tube running along their back. Maybe all this demonstrates is that there is more than one way to arrange a symmetrical body. But people who like to think that evolution is a devious process have suggested that the human and insect arrangements are actually derived from a single ancestor. If this organism had a nerve tube in its belly and a gut tube in its back like an insect, is it possible that some of this animal's descendants evolved into insects, but that some others turned over on to their backs and became humans?

This seems a rather bizarre suggestion – could we vertebrates really be upside-down insects? Indeed, for a long time this theory was neglected, or remembered mainly as an example of the follies of scientific overenthusiasm. Remarkably enough, however, modern developmental research has provided genetic evidence that actually supports this theory. Next time you see a fly hanging upside-down on the ceiling, just remember that he may actually be the same way up as your dog who is standing on the floor.

p. 151: Deciding what sex to be

For something as important as offspring sex, you would have thought that most animals would employ similar methods to programme their babies. In fact, nothing could be further from the truth. Different animal species employ a wide variety of mechanisms to choose the sex of their children.

The simplest method is to be hermaphrodite. Although this creates problems later on because hermaphrodite animals usually have to be careful not to impregnate themselves, life is simpler for the developing hermaphrodite embryo because it does not have to make any 'decision' at all about its sex. Hermaphroditism is rare in vertebrates, however, although it does occur in the jawless hagfish and is also surprisingly common in moles. These are very much exceptions – most animals prefer to come down on one side or the other of the gender fence.

Some vertebrates use their environment as a deciding factor in their offspring's sex. The best-known vertebrate examples of this phenomenon are fish (which may change gender according to social or visual cues) and reptiles (crocodiles and turtles often decide their gender according to the temperature of the sand in which their eggs are buried). This method obviously works well for many species, but for others, including humans, relying on weather is evidently a bit arbitrary and instead we devised genetic methods of gender determination.

Genetic sex determination can take many different forms. In some fish, for example, the system is arranged so that some embryos are diploid and some are haploid – the individuals with two sets of genetic material become females and those with a single set become male. However, most forms of genetic determination involve the inheritance of a specific genetic signal by embryos destined to be a particular sex. Sometimes, multiple genes cooperate to instil maleness or femaleness in embryos, but in other species, like humans, the presence or absence of a single gene dictates embryonic sex.

p. 152: X inactivation and tortoiseshell cats

A commonplace example of this phenomenon is the tortoiseshell or calico cat. In tortoiseshell cats, the orange fur colour gene is carried on the X chromosome. Female cats can be tabby or black if they are XX, or ginger if they are $X_{orange}X_{orange}$. However, if they happen to be $X_{orange}X$, then half of their fur will use the ginger gene and the other half will not, producing a ginger mosaic cat, a tortoiseshell. Male cats do not usually have this option – they are either ginger ($X_{orange}Y$) or not (XY). Rarely, however, male tortoiseshells are born, but they have an abnormal number of sex chromosomes, such as $XX_{orange}Y$, or are a mix of $X_{orange}X$ and XY cells.

Chapter 4: The visitor within

p. 171: The opposite of humans

Unlike most animals, humans have 'deciduate' placentas, where a considerable amount of the uterus is torn off and lost with the placenta at birth. Even less common is the 'contradeciduate' placenta seen in some species of mole. A contradeciduate placenta works the other way round: parts of the fetal membrane are torn off during birth and stay attached within the uterus.

p. 178: The genetics of blood transfusion

The blood proteins A and B are, in fact, inherited as dominant Mendelian traits. If the gene for the A protein is called A, and the recessive gene for no A protein is called a, and we use the same terminology for the B protein, the genes that decide what blood groups people have are as follows:

Blood group	Possible genetic make-up
O	*aabb*
A	*Aabb, AAbb*
B	*aaBb, aaBB*
AB	*AaBb, AABb, AaBB, AABB*

p. 182: Why is the MHC linked to disease susceptibility?

When scientists became suspicious that the MHC was of central importance in immunology, the first thing they did was to see if people's MHC genes have any effect on the diseases they get.

Unexpectedly, one set of diseases, the autoimmune diseases, showed strong links with certain MHC genes. Autoimmune diseases include a varied assortment of conditions, but they are all caused by the immune system itself. If the immune system is viewed as the security organisation of the body, then autoimmune disease occurs when it embarks on an ill-advised McCarthyite purge. Suddenly, the immune system fails to recognise some part of the body as 'self' and proceeds to try and destroy it. These misguided autoimmune purges are now known to cause rheumatoid arthritis, multiple sclerosis, coeliac disease, some forms of diabetes and thyroid disease as well as a variety of skin diseases.

The example that MHC buffs always quote is a disease called ankylosing spondylitis, in which the immune system attacks the joints. Ankylosing spondylitis is strongly linked to a single variant of MHC-B – around 90 per cent of people with this disease have the *MHC-B 27* gene. There are similar links between MHC types and other autoimmune diseases.

While it was interesting to find that susceptibility to many autoimmune diseases is dependent on MHC type, these diseases are really of peripheral importance in the history of the human race. Until this century, most people died of infectious diseases – the very conditions that modern medicine has been most effective at treating. So, presumably, the main function of our immune system for most of our history has been to repel infectious microbes. You might therefore expect a strong link between MHC molecules and susceptibility to infectious disease.

Disappointingly, it has proved difficult to demonstrate a relationship between infectious diseases and MHC genes. The only really good link is that between malaria and the some MHC genes prevalent in people from West Africa. These molecules appear to confer considerable resistance to malaria on the people who carry them. In fact, they are almost as protective as the sickle-cell haemoglobin gene mentioned in Chapter 1.

It is now thought that the links between MHC genes and disease are due to the abilities of the MHC molecules to present different protein fragments to the immune system. In the case of autoimmune disease, some MHC molecules may be more likely to mislead the immune system into thinking that some self-molecules are actually alien. This theory probably also explains the links between MHC molecules and infectious diseases. Presumably some MHC molecules present certain alien molecules better than others, and so they confer resistance to certain diseases. Why the link is so much more tenuous than that for autoimmune diseases is unknown.

p. 182: Why do MHC genes vary so much in the human population?

The apparently obsessive drive to keep a large number of different MHC genes in the population may be explained by the central role of the MHC in immunity. Probably, the need to maintain vigilance for a wide variety of different micro-organisms has encouraged populations to retain a wide array of MHC molecules. This ensures that most individuals will inherit different MHC genes from their two parents, so that they usually have two versions of each MHC gene. This may increase the range of microbial materials that their infected cells can present to their lymphocytes.

p. 192: Placental survival in other mammals

Fetuses' strategies to avoid attack by their mothers' immune systems differ greatly in other mammals. In the rat, for example, MHC genes inherited from the mother are switched off in the placenta. It is not known why this is so, especially as the MHC proteins made from the paternal genes are often prevented from reaching the surface of placental cells.

MHC expression on the placenta of horses and cattle is even more peculiar. Hoofed mammals have placentas in which the fetal cells are far less invasive than those found in humans. Despite this, a few cells still invade the mother's tissue. In the horse, for example, a set of fetal placental cells called the 'endometrial cups' ploughs deeply into the mare's uterine wall at around the seventh week of pregnancy. For some reason, these cells switch on their MHC genes just before they start to invade. Also, my co-workers and I have examined a completely different type of cell, the binucleate cell of the bovine placenta, which is also invasive and also switches on MHC just before its foray into maternal tissue. Just as in humans, it is the cells most exposed to the maternal immune system that switch on MHC, but horse and cattle invasive trophoblasts do not switch on a special placental gene, like human MHC-G. Instead, they switch on 'usual' MHC – the same molecules that cause transplantation reactions. We have not yet explained why these invasive cells should want to do such a dangerous thing.

p. 196: The risks of rhesus disease

Let us assume that 10 per cent of people are rhesus negative and 90 per cent of people are rhesus positive. This means that 10 per cent of women are rhesus negative and thus are potentially at risk of getting rhesus disease. Choosing at random, 90 per cent of these women would be expected to pick a rhesus-positive mate, meaning that 9 per cent of women are expected to be rhesus negative and to have selected a rhesus-positive mate (90 per cent of 10 per cent is 9 per cent).

Further reading

○ ○ ○ ○ ○ ○ ○ ○ ○ ○

Chapter 1: Origins

Like much of the information in this book, the sources of the information on the sperm's journey to the egg are scattered in different scientific journals. If you want to read more about how sperm reach their goal, you might want to track down

Baker, R. and R. Bellis. 1993. *Animal Behaviour* 46: 887.
Eisenbach, M. 1999. *Reviews of Reproduction* 4: 56.

The elegant experiments leading to the discovery of sperm–egg binding by Paul Wassarman and his co-workers are described in an article in *Scientific American*, always a good source of information for non-specialist readers.

Wassarman, P.M. 1988. *Scientific American* December: 52.

Gregor Mendel's experiments have been described in many biology texts, but good biographies are contained in

Bronowski, J. 1977. *Ascent of Man*. London: BBC.
Henig, R.M. 2000. *A Monk and Two Peas*. London: Weidenfeld & Nicolson.

For more about parthenogenesis, see

Grebelnyi, S.D. 1996. *Hydrobiologia* 320: 55.
Howard, R.S. and C.M. Lively. 1998. *Evolution* 52: 604.
Ineich, I. 1992. *Bulletin de la Société Zoologique de France – Evolution et Zoologie* 117: 253.
Lively, C.M. and S.G. Johnson. 1996. *Proceedings of the Royal Society of London B* 263: 1023.
MacCullogh, R.D., R.W. Murphy, L.A. Kupriyanova and I.S. Darevsky. 1997. *Biochemical Systematics and Ecology* 25: 33.
Moritz, C. et al. 1992. *Genetica* 87: 53.
Price, A.H. 1992. *Copeia* 2: 323.
Strain, L., J.P. Warner, T. Johnston, D.T. Bonthron. 1995. *Nature Genetics* 11: 164.

Taylor, A.S. and P.R. Braude. 1994. *Human Reproduction* 9: 2389.

Tinti, F. and V. Scali. 1995. *Evolution* 50: 1251.

Whittier, J.M., D. Stewart and L. Tolley. 1994. *Copeia* 2: 484.

The long-running debate on the reason for the size disparity between sperm and egg has included

Godelle, B. and X. Reboud. 1995. *Proceedings of the Royal Society of London B* 259: 27.

Hastings, I.M. 1992. *Genetic Research* 59: 215.

Hurst, L.D. and W.D. Hamilton. 1992. *Proceedings of the Royal Society of London B* 247: 189.

Levitan, D.R. 1996. *Nature* 382: 153.

On competition between sperm, see

Gomendio, M. and E.R. Roldan. 1991. *Proceedings of the Royal Society of London B* 243: 181.

Short, R.V. 1997. *Acta Paediatrica Supplement* 422: 3.

Further details of fertilisation and early embryonic development can be found in

Heikinheimo, O. and W.E. Gibbons. 1999. *Molecular Human Reproduction* 4: 745.

Henrion, G., A. Brunet, J.P. Reyard and V. Duranthon. 1997. *Molecular Reproduction and Development* 47: 353.

Howlett, S.K. and V.N. Bolton. 1985. *Journal of Embryology and Experimental Morphology* 87: 175.

Myles, D.G. 1993. *Developmental Biology* 158: 35.

Nothias, J.Y., M. Miranda and M. DePamphilis. 1996. *EMBO Journal* 15: 5715.

Pardee, A.B. 1989. *Science* 240: 603.

Parrish, J.J., C.I. Kim and I.H. Bae. 1992. *Theriogenology* 38: 277.

Stebbins-Boaz, B. and J.D. Richter. 1997. *Critical Reviews in Eukaryotic Gene Expression* 7: 73.

Tesarik, J., V. Kopechny, M. Plachot and J. Mandelbaum. 1987. *Development* 101: 777.

Wang, Q. and K.E. Latham. 1997. *Molecular Reproduction and Development* 47: 265.

Wassarman, P.M. 1987. *Science* 235: 553.

Xu, K.P. and W.A. King. 1990. *AgBiotech News and Information* 2: 25.

Fruit fly development is a large area of research and it would probably be possible to fill an entire book with just the titles of journal articles published about it, but for more about the maternal polarity gene *bicoid*, see

Frohnhofer, H.G. and C. Nusslein-Vollhard. 1986. *Nature* 324: 120.

For information on zebrafish and *Xenopus* maternal genes, see

Santacruz, H. and S. Vriz. 1996. *Biology of the Cell* 88: 153.

Satou, Y. and N. Satoh. 1997. *Developmental Biology* 192: 467.

Yoshida, S., Y. Marikawa and N. Satoh. 1996. *Development* 122: 2005.

Yoshida, S., Y. Satou and N. Satoh. 1997. *Cold Spring Harbor Symposia on Quantitative Biology* 62: 89.

Zhang, J. et al. 1998. *Cell* 4: 515.

Molecular biology and the central dogma is an even larger research area and I have been lucky to work with several people who have known a great deal about it. For a thorough grounding, try the following epic:

Watson, J.D., N.H. Hopkins, J.W. Roberts, J.A. Steitz and A.M. Wiener. 1987. *Molecular Biology of the Gene*. Menlo Park, CA: Benjamin/Cummings.

Chapter 2: Breaking the cycle

Obviously, the direct way to approach the works of William Harvey is to read them. Because of its central position in modern science, *De Motu Cordis* is relatively freely available. I have an 'Everyman's Library' copy published by J.M. Dent and Sons, 1990. Getting hold of *De Generatione Animalium* is rather more difficult; I borrowed a version published for the Sydenham Society in 1847 from the Wellcome Trust Library in Euston Road. Track it down if you can. I must admit, I find the long initial section about birds rather tough going, but the deer bits make up for the slog. For the purist, there may be Latin versions available too, although I have it on good authority that the original English translations capture the mood of Harvey's time rather better, and so the non-linguists among us need not worry. As well as the primary texts, I was also greatly aided by a biography of Harvey:

Chauvois, L. 1957. *William Harvey*. London: Hutchinson and Co.

For general reproduction and hormonal information, my favourite textbook is

Findlay, A.L.R. 1984. *Reproduction and the Fetus*. London: Edward Arnold.

and for specific papers about the hormones of the reproductive cycle, try

Campbell, B.K., G.E. Mann, A.S. McNeilly and D.T. Baird. 1990. *Endocrinology* 127: 227.

Caraty, A., A. Locatelli and G.B. Martin. 1989. *Journal of Endocrinology* 123: 375.

Denamur, R., J. Martinet and R.V. Short. 1973. *Journal of Reproduction and Fertility* 32: 207.

Espey, L.L. 1994. *Biology of Reproduction* 50: 233.

Karsch, F.J., D.L. Foster, E.L. Bittman and R.L. Goodman. 1983. *Endocrinology* 113: 1333.

Karsch, F.J., D.L. Foster, S.J. Legan, K.D. Ryan and G.K. Peter. 1979. *Endocrinology* 105: 421.

Karsch, F.J et al. 1984. *Recent Progress in Hormone Research* 40: 185.

Murdoch, W.J. 1995. *Biology of Reproduction* 53: 8.

O'Shea, J.D., D.G. Cran and M.F. Hay. 1980. *Cell and Tissue Research* 210: 305.

Scaramuzzi, R.J. et al. 1993. *Reproduction, Fertility and Development* 4: 459.

Smeaton, T.C. and H.A. 1971. *Journal of Reproduction and Fertility* 25: 243.

On why women can have sex all the time,

Diamond, J. 1997. *Why is Sex Fun?* London: Weidenfeld & Nicolson.

why they menstruate,

Finn, C.A. 1998. *Quarterly Review of Biology* 73: 163.
Giannini, A.J., W.A. Price and R.H. Loiselle. 1984. *International Journal of Psychophysiology* 1: 341.
Korzekwa, M.I. and M. Steiner. 1987. *Clinics in Obstetrics and Gynaecology* 40: 564.
Redei, E. and E.W. Freeman. 1993. *Acta Endocrinologica* 128: 536.
Strassmann, B.I. 1996. *Quarterly Review of Biology* 71: 181.

why menstruation protects them against hereditary haemochromatosis,

Little, P. 1996. *Nature* 382: 494.

and why endometriosis may be caused by pollution:

Rier, S.E., D.C. Martin, R.E. Bowman and J.L. Becker. 1995. *Environmental Health Perspectives* 103: 151.

For a general review of mammalian strategies for the maternal recognition of pregnancy, try the admirably clear

Flint, A.P.F, J.P. Hearn and A.E. Michael. 1990. *Journal of Zoology* 221: 327.

and here are some papers and books for more details. The information on deer maternal recognition of pregnancy is very similar to earlier work on sheep, so I have quoted some papers on them here too.

Bainbridge, D.R.J and H.N. Jabbour. 1997. *Journal of Reproduction and Fertility* 111: 299.
Bainbridge, D.R.J and H.N. Jabbour. 1999. *Journal of Reproduction and Fertility* 116: 305.
Bainbridge, D.R.J., M.H. Davies, R.J. Scaramuzzi and H.N. Jabbour. 1996. *Biology of Reproduction* 55: 883.
Christiansen, I.J. 1984. *Reproduction in the Dog and Cat*. London: Baillière Tindall.
Clutton-Brock, T.H., F.E. Guinness and S.D. Albon. 1992. *Red Deer*. Edinburgh: Edinburgh University Press.
Flint, A.P.F. 1995. *Reproduction, Fertility and Development* 7: 313.
Flint, A.P.F. and E.L. Sheldrick. 1983. *Journal of Reproduction and Fertility* 67: 125.
Flint, A.P.F., S.D. Albon, A.S. Loudon and H.N. Jabbour. 1997. *Hormones and Behaviour* 31: 296.
Gemmell, R.T. 1995. *Reproduction, Fertility and Development* 7:303.
Godkin, J.D., F.W. Bazer, J. Moffatt, F. Sessions and R.M. Roberts. 1982. *Journal of Reproduction and Fertility* 65: 141.

Guinness, F., G.A. Lincoln and R.V. Short. 1971. *Journal of Reproduction and Fertility* 27: 427.

Hayssen, V., R.C. Lacey and P.J. Parker. 1985. *American Naturalist* 126: 617.

Lillegraven, J.A., S.D. Thompson, B.K. McNab and I.L. Patton. 1987. *Biological Journal of the Linnean Society* 32: 281.

Mead, R.A. 1993. *Journal of Experimental Zoology* 266: 629.

Renfree, M.B. 1993. *Journal of Experimental Zoology* 266: 450.

Short, R.V. 1969. *Ciba Foundation Symposia* 111: 2.

Silvia, W.J., G.S. Lewis, J.A. McCracken, W.W. Thatcher and L. Wilson. 1991. *Biology of Reproduction* 45: 655.

Vallet, J.L., G.E. Lamming and M. Batten. 1990. *Reproduction, Fertility and Development* 90: 265.

The information on morning sickness comes from

Crystal, S.R. and I.L Bernstein. 1998. *Appetite* 30: 297.

Goodwin, T.M., M. Montoro, J.H. Mestman, A.E. Pekary and J.M. Hershman. 1992. *Journal of Clinical Endocrinology and Metabolism* 75: 1333.

Yoshimura, M. and J.M. Hershman. 1995. *Thyroid* 5: 425.

The book that started the morning-sickness-as-protective-mechanism debate is

Profet, M. 1997. *Protecting your Baby-to-be*. New York: Warner.

Chapter 3: Making babies

Ernst Haeckel wrote in a bewilderingly turgid way. If Darwin had trouble ploughing through his books, then it is a fairly safe assumption that you and I would not have much luck either. Perhaps the best one to read and one *is* the right number to read, in my opinion, is the modestly titled *Riddle of the Universe at the Close of the Nineteenth Century*. In this slim volume Haeckel, in a rather *fin de siècle* mood, reviews life, the Universe and everything and even has time to expound his religious beliefs before tea. It is reasonably easy to get hold of. I have a copy from the 'Great Minds' series published by Prometheus Books, Buffalo, NY.

Haeckel, E.H.P.A. 1867. *General Morphology of Organisms*. Berlin: Georg Reimer.

Haeckel, E.H.P.A. 1900. *The Riddle of the Universe at the Close of the Nineteenth Century*. New York: Harper and Brothers.

Of course, Darwin is the overarching character in all this, and so it would be wrong to leave out *that* book as well his other magnum opus and an excellent biography.

Darwin, C. 1859. *On the Origin of Species by Means of Natural Selection*. London: John Murray.

Darwin, C. 1871. *The Descent of Man, and Selection in Relation to Sex*. London: John Murray.

Desmond, A. and J. Moore. 1991. *Darwin*. London: Michael Joseph.

The scientific debate over the links between ontogeny and phylogeny started long before Haeckel and is still not over:

von Baer, K.E. 1828. *Entwickelungsgeschichte der Thiere: Beobachtung und Reflexion*. Berlin: Bornträger.

Berge, C. 1998. *American Journal of Physical Anthropology* 105: 441.

Garstang, W. 1922. *Zoological Journal of the Linnean Society* 35: 81.

Gee, H. 1996. *Before the Backbone*. London: Chapman and Hall.

Gould, S.J. 1977. *Ontogeny and Phylogeny*. Cambridge, MA: Harvard University Press.

Gould, S.J. 1990. *Wonderful Life*. London: Hutchinson Radius.

On developmental genes shared between flies and mammals, see

Akam, M. 1988. *Cell* 5: 347.

Graham, A., N. Papalopulu and R. Krumlauf. 1989. *Cell* 5: 367.

The exposure of Haeckel's economy with the *actualité* of vertebrate embryology has led to a discussion of the relation between observation, representation and interpretation in science that has ranged from humorous through sanctimonious to esoteric.

Kemp, M. 1998. *Nature* 395: 447.

Pennisi, E. 1997. *Science* 277: 1435.

Richardson, M.K. 1998. *Science* 281: 1289.

Richardson, M.K. et al. 1997. *Anatomy and Embryology* 196: 91.

Richardson, M.K. et al. 1998. *Science* 280: 983.

If you want a more detailed insight into embryology, try

Gilbert, S.F. and A.M. Raunio. 1997. *Embryology: Building the Organism*. Sunderland, MA: Sinauer.

I have subdivided the rest of further reading for descriptive embryology into 'general and pre-organ systems embryology',

De Robertis, E.M. and Y. Sasai. 1996. *Nature* 380: 37.

Dickmann, Z. 1969. *Advances in Reproductive Physiology* 4: 187.

Johnson, M.H. 1996. *Reproduction, Fertility and Development* 8: 699.

Keith, L. and G. Machin. 1997. *Journal of Reproductive Medicine* 42: 699.

Lummaa, V., E. Haukioja, R. Lemmetyinen and M. Pikkola. 1998. *Nature* 294: 533.

Romer, A.S. 1933. *Man and the Vertebrates II*. Harmondsworth: Pelican.

'organ system embryology',

Ahlberg, P.E. 1997. *Nature* 85: 489.

Kawamura, K. and S. Kikuyama. 1998. *Archives of Histology and Cytology* 61: 189.

Köntges, G. and A. Lumsden. 1996. *Development* 122: 3229.

Kuratani, S. 1997. *Anatomy and Embryology* 195: 1.

Merino, R., Y. Ganan, D. Macias, J. Rodriguez-Leon and J.M. Hurle. 1999. *Annals of the New York Academy of Science* 887: 120.
Northcutt, R.G. 1993. *Acta Anatomica Basel* 148: 71.
Riji, F.M., A. Gavalas and P. Chambon. 1998. *International Journal of Developmental Biology* 42: 393.
Wake, D.B. 1993. *Acta Anatomica Basel* 148: 124.

and 'embryology of the reproductive system':

Haqq C.M. et al. 1994. *Science* 266: 1494.
Jegalian, K. and D.C. Page. 1998. *Nature* 394: 776.
Leaman, T., R. Rowland and S.E. Long. 1999. *Veterinary Record* 144: 9.
Mittwoch, U. 1993. *Human Reproduction* 8: 1550.
Simpson, E., D. Scott and P. Chandler. 1997. *Annual Review of Immunology* 15: 39.
Werren, J.H. 1998. *Annual Review of Ecology and Systematics* 29: 233.

For more about hydatidiform moles,

Jacobs, P.A., C.M. Wilson, J.A. Sprenkle, N.B. Rosenshein and B.R. Migeon. 1980. *Nature* 286: 714.
Ohama, K. et al. 1981. *Nature* 291: 551.

the two-headed boy of Bengal,

Bondeson, J. 1997. *A Cabinet of Medical Curiosities.* Ithaca, NY: Cornell University Press.

the strange story of fetus-in-fetu,

Anon. 1996. *Fortean Times* 85: 18.
Hopkins, K.L. et al. 1992. *Journal of Paediatric Surgery* 32: 1476.

the Burgess shales,

Whittington, H.B. 1985. *The Burgess Shale.* Newhaven, CT: Yale University Press.

and fetal hearing:

Deliege, I. and J. Sloboda. 1996. *Musical Beginnings.* Oxford: Oxford University Press.
Hepper, P.G. 1991. *Irish Journal of Psychology.* 13: 95.
Rauscher, F.H., G.L. Shaw amd K.N. Ky. 1993. *Nature* 361: 611.
Shahidullah, S. and P.G. Hepper. 1992. *International Journal of Prenatal and Perinatal Studies* 4: 235.

Here are two references about 'treatment' of intersexuality. This is a very controversial field, and for every scholarly assertion there is usually an equally scholarly contradiction elsewhere.

Kipnins, K. and M. Diamond. 1998. *Journal of Clinical Ethics* 9: 398.

Money, J. and A. Ehrhardt. 1972. *Man and Woman, Boy and Girl*. Baltimore: Johns Hopkins University Press.

Chapter 4: The visitor within

My favourite book on the strange quirks of the placenta is by Donald Steven, head of the Veterinary Anatomy Department at Cambridge when I trained there. It is called *Comparative Placentation* and was published by Academic Press in 1975. For other information on placental structure and function, try

Benirschke, K. and P. Kaufmann. 1990. *Pathology of the Human Placenta*. Berlin: Springer-Verlag.
Llewellyn-Jones, D. 1990. *Fundamentals of Obstetrics and Gynaecology*. London: Faber and Faber.
Mi, S. et al. 2000. *Nature* 403: 785.

For the story of the *ari-ari*, read Chapter 10 of

Eiseman, F.B. 1990. *Bali: Sekala and Niskala*. Hong Kong: Periplus Editions.

Just like genetics, hormones and embryology, immunology is now an enormous discipline that we have just touched upon. To learn more about how animals protect themselves, you might want to try one of the following. The first is a good 'scene-setter' if you are interested in the evolution of the immune system:

Beck, G. and G.S. Habicht. 1996. *Scientific American* November: 42.
Janeway, C. and P. Travers. 1997. *Immunobiology*. New York: Garland.
Kuss, R. and P. Bourget. 1992. *An Illustrated History of Transplantation*. Rueil-Malmaison: Laboratoires Sandoz.
Roitt, I.M. 1994. *Essential Immunology*. Oxford: Blackwell.

For some details on the MHC, try

Browning, M. and A.J. McMichael. 1996. *HLA and MHC: Genes, Molecules and Function*. London: Academic Press (especially Chapter 1 by Jim Kaufman).
Hill, A.V.S. et al. 1995. *Nature* 352: 595.
Potts, W.K., C.J. Manning and E.K. Wakeland. 1991. *Nature* 352: 619.
Yeager, M. and A.L. Hughes. 1999. *Immunological Reviews* 167: 45.

The sexy smelly T-shirt experiment is described in

Wedekind, C. 1994. *Philosophical Transactions of the Royal Society of London B* 346: 303.

On the link between MHC and fertility, see

Beer, A.E., J.F. Quebbeman, J.W.T. Ayers and R.F. Haines. 1981. *American Journal of Obstetrics and Gynaecology* 141: 987.
Ho, H.N. et al. 1994. *American Journal of Obstetrics and Gynaecology* 170: 63.

Komlos, L., R. Zamir, H. Joshua and I. Halbrecht. 1977. *Clinical Immunology and Immunopathology* 7: 330.

Ober, C. 1992. *Experimental and Clinical Immunogenetics* 9: 1.

Peter Medawar's seminal paper 'Some immunological and endocrinological problems raised by the evolution of viviparity in vertebrates' was published as

Medawar, P.B. 1953. *Symposia of the Society for Experimental Biology* 7: 320.

Here are some pregnancy immunology references. For general reading, try

Loke, Y.W. and A. King. 1995. *Human Implantation*. Cambridge: Cambridge University Press.

Redman, C.W.G., I.L. Sargent and P.M. Starkey. 1993. *The Human Placenta*. Oxford: Blackwell.

or for specific information about genetic imprinting,

Bartolomei, M.S. and S.M. Tilghman. 1997. *Annual Review of Genetics* 31: 493.

Itier, J.M. et al. 1998. *Nature* 393: 125.

Leighton, P.A., J.R. Saam, R.S. Ingram and S.M. Tilghman. 1996. *Biology of Reproduction* 54: 273.

Nicholls, R.D., S. Saitoh and B. Horsthemke. 1998. *Trends in Genetics* 14: 194.

placental non-imprinting of MHC antigens,

Bainbridge, D.R.J., S.A. Ellis and I.L. Sargent. 1999. *Journal of Immunology* 163: 2023.

Hashimoto, K. et al. 1997. *Japanese Journal of Human Genetics* 42: 181.

placental MHC in humans,

Braude, V.M., D.S.J. Allen, D. Wilson and A.J. McMichael. 1997. *Current Biology* 8: 1.

Chumbley, G., A. King, N. Holmes and Y.W. Loke. 1993. *Human Immunology* 37: 17.

Ellis, S.A, M.S. Palmer and A.J. McMichael. 1990. *Journal of Immunology* 144: 731.

Ellis, S.A., I.L. Sargent, C.W.G. Redman and A.J. McMichael. 1986. *Immunology* 59: 595.

Geraghty, D.E., B.H. Koller and H.T Orr. 1987. *Proceedings of the National Academy of Sciences USA* 84: 9145.

Ishitani, A. and D.E. Geraghty. 1992. *Proceedings of the National Academy of Sciences USA* 89: 3947.

King, A. et al. 1996. *Journal of Immunology* 156:2068.

Kovats, S. et al. 1990. *Science* 248: 220.

Lee, N.A.R. et al. 1995. *Immunity* 3: 591.

Pazmany, L. et al. 1996. *Science* 274: 794.

Redman, C.W.G., A.J. McMichael, G.M. Stirrat, C.A. Sunderland and A. Ting. 1984. *Immunology* 52: 457.

Sanders, S.K., P.A. Giblin and P. Kavathas. 1991. *Journal of Experimental Medicine* 174: 737.

Starkey, P.M. 1987. *Journal of Reproductive Immunology* 11: 63.

Sunderland, C.A., M. Naiern, D.Y. Mason, C.W.G. Redman and G.M. Stirrat. 1981. *Journal of Reproductive Immunology* 3: 323.

Watkins, D.I. et al. 1990. *Nature* 346: 60.

Yelavarthi, K.K., J.L. Fishback and J.S. Hunt. 1991. *Journal of Immunology* 146: 2487.

and other animals:

Bainbridge, D.R.J., I.L. Sargent and S.A. Ellis. 1999. *Placenta* 20: A7.

Donaldson, W.L., C.H. Zhang, J.G. Oriol and D.F.Antczak. 1990. *Development* 110: 63.

Ellis, S.A., B. Charleston, I.L. Sargent and D.R.J. Bainbridge. 1998. *Journal of Reproductive Immunology* 37: 103.

Beyond the 'placenta as graft', read

Abderhalden, E. 1912. *Die Schutzfermente*. Berlin: Springer.

Aït-Azzouzene, B. 1998. *Journal of Immunology* 161: 2673.

Antczak, D.F. 1989. *Current Opinion in Immunology* 1: 1135.

Bainbridge, D.R.J., S.A. Ellis and I.L. Sargent. 2000. *Journal of Reproductive Immunology* 47: 1.

Chua, S., T. Wilkins, I.L. Sargent and C.W.G. Redman. 1991. *British Journal of Obstetrics and Gynaecology* 98: 973.

Hunt, J.S. 1991. *Current Opinion in Immunology* 4: 591.

Munn D.H. 1998. *Science* 281: 1191.

Sacks, G., I.L. Sargent and C.W.G. Redman. 1999. *Immunology Today* 20: 114.

Wegmann, T.G., H. Lin, L. Guilbert and T.R. Mosmann. 1993. *Immunology Today* 14: 353.

The 'Mummy made me gay' paper is

Blanchard, R. and P. Klassen. 1997. *Journal of Theoretical Biology* 185: 373.

Here are two papers about cord blood transplantation:

Gordon, M.Y., J.L. Lewis, M.A. Scott, I.A. Roberts and J.M. Goldman. 1995. *British Journal of Haematology* 90: 744.

Hows, J.M. et al. 1992. *Lancet* 340: 73.

Chapter 5: The visitor without

It seems rather disconcerting to strip down something like infant death to dry statistical trends and impersonal health strategies, but of course that is what many scientists and medics spend their time doing. I could not find many general reviews of the history and geography of perinatal mortality, and so I pieced my discussion of it together from some of the following papers.

Aleman, J., J. Liljestrand, R. Pena, S. Wall and L.A. Persson. 1997. *Gynecologic and Obstetric Investigation* 43: 112.

Arntzen, A., T. Moum, P. Magnus and L.S. Bakketeig. 1996. *Scandinavian Journal of Social Medicine* 24: 36.

Avery, M.E. 1992. *Early Human Development* 29: 43.

Beebe, S.A., J.R. Britton, H.L. Britton, P. Han and B. Jepson. 1996. *Pediatrics* 98: 231.

Bottoms, S.F. et al. 1997. *British Medical Journal* 314: 1521.

Cnattingius, S. and M.L. Nordstrom. 1996. *Acta Paediatrica* 85: 1400.

DeLong, G.R. et al. 1997. *Lancet* 350: 771.

Kossel H. and H. Versmold. 1997. *Journal of Perinatal Medicine* 25: 421.

Moore, R.F. 1996. *Ethology and Sociobiology* 17: 379.

Nielsen, B.B., J. Liljestrand, M. Medegaard, S.H. Thilsted and A. Joseph. 1997. *British Medical Journal* 314: 1521.

Stoll, B.J. 1997. *Clinics in Perinatology* 24: 1.

Taha, T.E. et al. 1997. *British Medical Journal* 315: 7102.

Papers on parental fears as birth approaches,

Almond, B.R. 1998. *International Journal of Psychoanalysis* 79: 775.

Areskog, B., N. Uddenberg and B. Kjessler. 1984. *Journal of Psychosomatic Research* 28: 213.

Szeverenyi, P., R. Poka, M. Hetey and Z. Torok. 1998. *Journal of Psychosomatic Obstetrics and Gynaecology* 19: 38.

couvade,

Bogren, L.Y. 1985. *Acta Psychiatrica Scandinavica* 71: 311.

Masoni, S., A. Maio, G. Trimarchi, C. de Punzio and P. Fioretti. 1994. *Journal of Psychosomatic Obstetrics and Gynaecology* 15: 125.

Tenyi, T., M. Trixler and F. Jadi. 1996. *Psychopathology* 29: 252.

and the story of the women who gave birth to rabbits and 365 children (two different women!):

Bondeson, J. 1997. *A Cabinet of Medical Curiosities*. Ithaca, NY: Cornell University Press.

Bondeson, J. 1997. *Fortean Times* 105: 40.

On the plains viscacha, see

Weir, B.J. 1971. *Journal of Reproduction and Fertility* 25: 355.

For some publications on the birth process, try the list below.

Bravo, P.W. 1994. *Veterinary Clinics of North America: Food Animal Practice* 10: 265.

Detrick, J.M., J.W. Pearson and R.C. Frederickson. 1985. *Obstetrics and Gynaecology* 65: 647.

Enkin et al. 1989 *A Guide to Effective Care in Pregnancy and Childbirth*. Oxford: Oxford University Press.

Gentry, P.A. and R.M. Liptrap. 1988. *Canadian Journal of Physiology and Pharmacology* 66: 671.

McLean, M. et al. 1995. *Nature Medicine* 1: 460.

Nathanielz, P.W. 1998. *European Journal of Obstetrics, Gynaecology and Reproductive Biology* 78: 127.

Smith, R. 1999. *Scientific American* March: 50.

Smith, R., S. Mesiano, E.C. Chan, S. Brown and R.B. Jaffe. 1998. *Journal of Clinical Endocrinology and Metabolism* 83: 2916.

Whipple, B., J.B. Josimovich and B.R. Komisaruk. 1990. *International Journal of Nursing Studies* 27: 213.

On selective amnesia during childbirth, see

Niven, C.A. and E.E. Brodie. 1996. *Pain* 64: 387.

Norvell, R.T., F. Gaston-Johansson and G. Fridh. 1987. *Pain* 31: 77.

Oujevolk, A.V., C.A. Christman and J. Altrocchi. 1995. *Journal of the American Board of Family Practitioners* 8: 410.

The vexed question of whether it is safer to have babies at home is considered by

Campbell, R. and A. MacFarlane. 1986. *British Journal of Obstetrics and Gynaecology* 93: 675.

Parazzini, F. and C. La Vecchia. 1988. *American Journal of Public Health* 78: 706.

Tew, M. 1979. *Lancet* 1: 1388.

Here is some further reading on the adaptations of the baby to its new environment. The strange story of the Julia Creek dunnart is the second paper on the list.

Mitchell, B.F., X. Fang and S. Wong. 1998. *Reviews of Reproduction* 3: 113.

Mortola, J.P., P.B. Frappell and P.A. Woolley. 1999. *Nature* 397: 660.

Rigatto, H. 1996. *Reproduction Fertility and Development* 8: 23.

Vialet, R. et al. 1996. *Annales de Pediatrie* 43: 99.

For information about Mother's adaptations to her new predicament, as well as some papers about the menopause, see the following:

Carter, C.S. and M. Altemus. 1997. *Annals of the New York Academy of Sciences* 807: 164.

Forsyth, I.A. 1986. *Journal of Dairy Science* 69: 886.

Hayssen, V. 1993. *Journal of Dairy Science* 76: 3213.

Kent, G.N. et al. 1990. *Journal of Bone Mineral Research* 5: 361.

Mepham, B. 1976. *The Secretion of Milk*. London: Edward Arnold.

Packer, C., M. Tatar and A. Collins. 1998. *Nature* 392: 807.

Peaker, M. 1977. *Comparative Aspects of Lactation*. London: Academic Press.

Peccei, J.S. 1995. *Maturitas* 21: 83.

Prentice, A.M., G.R. Goldberg and A. Prentice. 1994. *European Journal of Clinical Nutrition* 48: suppl. 3: 78.

Westendorp, R.G.H and T.B.L. Kirkwood. 1998. *Nature* 396: 743.

Whitworth, N.S. 1988. *Psychoneuroendocrinology* 13: 171.

For the advantages of breastfeeding, try

Golding, J., P.M. Emmett and I.S. Rogers. 1997. *Early Human Development* 29: s131.

Jacobson, S.W., L.M. Chiodo and J.L. Jacobson. 1999. *Pediatrics* 103: e 71.

Koutras, A.K. and V.J. Vigorita. 1989. *Journal of Paediatric Gastroenterology and Nutrition* 9: 58.

Lucas, A., R. Morley, T.J.Cole, G. Lister and C. Leeson-Payne. 1992. *Lancet* 339: 261.

Paine, R. and R.J. Coble. 1982. *American Journal of Diseases of Children* 136: 36.

Riva, E. et al. 1996. *Acta Paediatrica* 85: 56.

Wigg, N.R. et al. 1998. *Australian and New Zealand Journal of Public Health* 22: 232.

Postnatal mood disorders are described in

Cox, J.L., J.M. Holden and R. Sagovsky. 1987. *British Journal of Psychiatry* 150: 872.

Murray, L. et al. 1999. *Journal of Child Psychology and Psychiatry* 40: 1259.

Pfuhlmann, B., E. Franzek, H. Beckmann and G. Stober. 1999. *Psychopathology* 32: 192.

For information about new fertility techniques, try the following. The first is about animals, and the third about people. The second is an interesting social history of human infertility.

Bainbridge, D.R.J. and H.N. Jabbour. 1998. *Veterinary Record* 143: 159.

Morice, P., P. Josset, C. Chapron and J.B. Dubuisson. 1995. *Human Reproduction Update* 1: 497.

Yovich J. and G. Grudzinskas. 1990. *The Management of Infertility*. Oxford: Heinemann.

Finally, here are some attempts at ectogenesis:

Kuwabara, Y. et al. 1987. *Artificial Organs* 11: 224.

Priscott, P.K., G.C. Yeoh and I.T. Oliver. 1984. *Journal of Experimental Zoology* 230: 247.

Sakata, M., K. Hisano, M. Okada and M. Yasufuku. 1998. *Journal of Thoracic and Cardiovascular Surgery* 115: 1023.

Unno, N. et al. 1993. *Artificial Organs* 17: 996.

Unno, N. et al. 1997. *Artificial Organs* 21: 1239.

Unno, N. et al. 1998. *Paediatric Research* 43: 452.

Yasufuku, M., K. Hisano, M. Sakata and M. Okada. 1998. *Journal of Paediatric Surgery* 33: 442.

Glossary

∘ ∘ ∘ ∘ ∘ ∘ ∘ ∘ ∘

As with most glossaries, this one contains brief explanations of some of the terms used repeatedly in this book. However, some entries are simply intended to provide the reader with the technical jargon terms for some of the simplified names I have used. I include these technical terms because they are used in most of the 'Further reading' I have recommended.

Acquired immune deficiency syndrome (AIDS) A disease caused by the Human Immunodeficiency Virus (HIV) attacking *thymic helper lymphocytes* and brain cells. The main result is a dramatic, but delayed, suppression of the immune system.

ACTH Adrenocorticotrophic hormone.

AIDS See *acquired immune deficiency syndrome*.

Allantois A *fetal membrane*, actually a sac derived from the early *endoderm*, much of which protrudes from the navel. The portion inside the fetus forms most of the bladder.

Amnion A *fetal membrane*, actually a sac derived from the early *ectoderm*. In humans, the amnion comes to surround the fetus almost completely. It lies inside the *trophoblast*.

Androgens A group of related steroid hormones. Most androgens in men are produced by the testicles. Androgens participate in the development of the male body and the production of sperm.

Anisogamy Sexual reproduction in which one sex produces sex cells much larger than those of the other. Cf. *isogamy*.

Antibody lymphocyte A cell of the immune system that makes antibodies: proteins designed to attach to, and precipitate the demise of, foreign organisms.

Aorta The largest artery in the body, it leaves the left side of the heart and carries blood to all the body except the lungs.

Artery A blood vessel that carries blood away from the heart. Cf. *vein*.

Asexual Reproduction without sex, such as *binary fission*, budding or *parthenogenesis*.

Bacterium A single-celled organism that does not have a separate *nucleus* to contain its genes.

Ball-of-cells (before blastocyst) Morula.

277

Bicoid A gene that encodes a product which a female fruit fly inserts into one end of her eggs, and which confers a front-to-back orientation on the developing embryo.

Bilirubin A yellow by-product of the destruction of *haemoglobin* which is normally excreted by the liver into the feces. If excess bilirubin is produced, or its excretion is blocked, it can accumulate in the skin, causing jaundice.

Binary fission A method of reproduction in which an animal simply splits into two offspring.

Biogenetic law A theory which states that embryos pass through all the stages of their evolution as they develop to adulthood. Thus embryonic development 'recapitulates' evolution.

Blastocyst An embryo that has differentiated into a hollow ball of cells (*trophoblast*) containing within it the *inner cell mass*.

Bone block Sclerotome.

Cellular trophoblast (cytotrophoblast) Originally the inner layer of the developing human trophoblast, this tissue later carries out much of the invasion into maternal tissues. Cf. *fused trophoblast*.

Chimaera An animal produced by a mixture of dissimilar cells. Cf. *hybrid*.

Chromosome A distinct structure within the *nucleus* of a *eukaryotic* cell which contains a portion of that cell's *genes* encoded on a single *DNA* molecule.

Clone A set of organisms generated *asexually* from a single ancestor.

Conception The process of becoming pregnant. I like to use this as a term to include all the different mechanisms that must act for a pregnancy to be established, of which fertilisation is only one.

Corpus luteum A temporary structure on the ovary which secretes *progesterone*.

Cortisol A steroid hormone produced by the adrenal glands. It has diverse effects around the body, but it is involved in fetal lung maturation in sheep and humans, and the initiation of labour in sheep.

Cretinism A congenital condition resulting from underactivity of the thyroid gland during fetal life.

CRH Corticotrophin releasing hormone.

Cumulus The diffuse cloud of cells which surrounds the mammalian *zona pellucida* and egg.

Cytoplasm The contents of a cell, other than the *nucleus*.

Decidua The thickened lining of the human uterus which is formed and shed every month. After fertilisation, the decidua is the tissue in which the embryo normally implants.

Deoxyribonucleic acid (DNA) A long linear molecule that carries along its length the digital codes which constitute the *genes*. In *eukaryotes*, the DNA is arranged into several distinct *chromosomes* contained in the *nucleus*.

Dermal armour Dermal bone.

DHEA Dihydroepiandrostenedione.

Diaphragm/liver bar Septum transversum.

DNA See *deoxyribonucleic acid*.

Dominant gene A gene inherited from one parent which can completely mask the presence of a counterpart *recessive gene* inherited from the other parent.

Thus, it needs to be inherited from only one parent for its effects to become apparent.

Double-egg twins Also known as 'fraternal' or, more correctly, 'dizygotic' twins, these twins are derived from two different eggs.

Down's syndrome This syndrome occurs because a baby inherits two copies of a certain chromosome from its mother instead of one. The syndrome is characterised by a typical facial and body shape, heart abnormalities and reduced life expectancy.

Duodenum The part of the small intestine that exits from the stomach and receives the tubes draining the pancreas and liver.

Early ectoderm Epiblast.

Early endoderm Hypoblast.

Ear plaque/bubble Otic placode/vesicle.

Ectoderm A layer of cells that forms on the top of the early embryo, and which goes on to form the amnion, the nervous system and the outer layer of the skin. Cf. *mesoderm, endoderm.*

Ectogenesis Growing a fetus outside its mother's body to a point at which it can survive in the outside world.

Ectopic pregnancy A pregnancy that implants outside the uterus.

Egg An egg is the germ cell made by a female animal, and from the scientific point of view, it ceases to be an egg once it is fertilised by a sperm. This is, of course, at odds with the 'everyday' definition of an egg: the egg laid by a hen is no longer a 'biological' egg if it has been fertilised and produces a chick.

Embryo A developing offspring during the period when most of its internal organs are forming (before eight weeks in humans). Cf. *fetus.*

Endocrine glands 'Internal' glands that release *hormones* into the blood.

Endoderm A layer of cells that forms on the bottom of the early embryo, and which folds in upon itself to form the gut, lungs, *yolk sac* and *allantois*. Cf. *mesoderm, ectoderm.*

Esophagus (gullet) The part of the gut between the pharynx and the stomach. In mammals the esophagus lies in the chest.

Estrogen A group of related steroid hormones. Estrogen in women is produced by the *Graafian follicles* and the *placenta*, if one is present. Estrogen participates in the development of the female body, the menstrual cycle, pregnancy and *lactation*.

Estrus (heat) A restricted period during which a female animal permits male animals to mate her.

Eukaryote An organism made up of cells with distinct *nuclei*. Cf. *bacterium.*

Evolution The gradual change of living species over time.

Eye plaque Lens placode.

Fetal membranes The sacs that surround the developing embryo and fetus. In different species, they include the *trophoblast, yolk sac, amnion* and *allantois*.

Fetus A developing offspring after the time at which most of its internal organs have formed (after eight weeks in humans). Cf. *embryo*. The word 'fetus' is derived directly from the Latin word *fetus*, which means 'offspring'. The alternative 'foetus' is a Middle English corruption of the same Latin word.

Final-kidney Metanephros.

First-kidney Pronephros.

Foetus See *fetus.*

Fused trophoblast (syncitiotrophoblast) Originally the outer layer of the developing human trophoblast, this tissue carries out the initial phase of invasion into maternal tissues. Cf. *cellular trophoblast.*

Gene The physical entity by which parents pass on characteristics to their offspring. Genes are instructions written as a digital code along the length of a *DNA* molecule.

Gill bars Branchial arches.

Gill grooves Branchial pouches (internal) and clefts (external).

Gonadotrophins Protein hormones produced by the *pituitary* which maintain activity of the ovaries or testicles.

Graafian follicles Bubble-like structures on the ovary, each of which contains a single developing egg. They rupture at ovulation.

Guard cells A catch-all term I have used for cells of the immune system that do not 'remember' previous infections, like *lymphocytes* do. They are a heterogeneous bunch, including macrophages, granulocytes, mast cells and natural killer cells.

Haemoglobin The red blood pigment, a protein that carries oxygen from the lungs or placenta to the tissues of the body.

HCG See *human chorionic gonadotrophin.*

Heat See *estrus.*

Hormones Chemical messenger molecules produced by one tissue (an *endocrine gland*) which enter the blood and have specific effects on distant tissues.

Human chorionic gonadotrophin A protein produced by the human embryo which prevents *luteolysis* and thus allows *maternal recognition of pregnancy* to take place.

Hybrid The production of offspring by breeding between two different species or strains. Cf. *chimaera.*

Imprinting The preferential use or disuse of a gene depending on the parent from which it was inherited.

Induced ovulation Ovulation in response to the act of copulation.

Inner cell mass A population of cells that forms inside the *trophoblast* when the embryo reaches the *blastocyst* stage. The inner cell mass is destined to form the baby, all the *fetal membranes* except the outermost one, and the blood vessels and connective tissue of the *placenta.*

Insulin A protein hormone produced by the pancreas. Although its main function is to control metabolism of sugars and fats, it also plays a role in maintaining *lactation.*

Interferon A group of proteins involved in suppressing viral infection, communication between cells of the immune system, and the *maternal recognition of pregnancy* in some hoofed mammals.

Intersex An individual who is neither an XX female nor an XY male. Intersexes can be chromosomal (with an abnormal sex chromosome content), gonadal (an apparently normal set of sex chromosomes fails to produce testicles or ovaries ['gonads'] as expected) or hormonal (apparently normal testicles or ovaries fail to cause the rest of the reproductive system to form normally).

In vitro fertilisation The generation of a zygote under artificial conditions by mixing eggs with sperm.

Isogamy Sexual reproduction in which both sexes produce sex cells of similar sizes. Cf. *anisogamy.*

Lactation The production of milk.

Larva The immature form of an animal which must undergo some radical rearrangement to become the adult form (metamorphosis).

Luteolysis Destruction of the *corpus luteum.*

Lymphocyte A cell of the immune system that responds to a very restricted set of alien materials, and which can be remain alive for many years after an infection to allow an animal to respond to the same infection should it arise again. I have crudely divided lymphocytes into *antibody, thymic helper* and *thymic killer* groups.

Major histocompatibility complex (MHC) An array of *genes* clustered together on a single *chromosome* which encodes a set of *proteins* that are central to transplantation acceptance or rejection. They are probably also important in several aspects of reproduction.

Maternal recognition of pregnancy The alteration of a mother's biology in response to the presence of a developing embryo, or embryos.

Meconium Tarry green feces accumulated in the gut during fetal life. Made from shed gut cells and digested amniotic fluid, it is passed in the first few days after birth.

Menopause The cessation of menstruation.

Mesoderm A layer of cells that forms in the middle of the early embryo, and which goes on to form most of the fetus. Cf. *ectoderm, endoderm.*

MHC See *Major histocompatibility complex.*

Mitochondria Structures within a *eukaryotic cell*'s *cytoplasm* responsible for reacting sugars with oxygen.

Mouth plate Prochordal plate, later buccopharyngeal membrane.

Müllerian ducts Folds in the *mesoderm* near the kidneys which can go on to form the Fallopian tubes, the uterus and the cervix.

Müllerian-inhibiting substance A substance produced by the testicles which suppresses the development of the *Müllerian ducts.*

Muscle block Myotome.

Mutation An alteration to an organism's genes, which may be passed on to its offspring.

Natural selection Darwin's proposed mechanism for *evolution,* which is based upon the diversity present in most natural *species.* Individuals with certain characteristics may be more successful, and so can pass on the genes for these characteristics to more offspring. Thus, the genetic make-up of entire species can gradually change over the generations.

Neural tube The early tubular precursor of the nervous system, formed from *ectoderm* under the influence of the *notochord.* The nervous system retains its basic tubular plan into adulthood.

Nose plaque Nasal placode.

Notch in early ectoderm Primitive streak.

Notochord A rod that forms in the embryonic *mesoderm* and which establishes the front-to-back orientation of vertebrate embryos. It also initiates the formation of the nervous system, the skeleton and most muscles.

Nucleus A separate compartment of a cell containing genes. Cf. *eukaryote*, *cytoplasm*.

Oxytocin A protein hormone secreted by the *pituitary* which is involved in different species in the destruction of the *corpus luteum*, the process of labour, milk secretion, parental bonding and the shrinking of the uterus after birth.

Paedomorphosis The retention of infantile, fetal or embryonic characteristics into the adult.

Parthenogenesis Production of offspring by an anatomically female animal without any contribution from a male.

Pharynx A widening of the front end of the vertebrate gut used variously for filter-feeding, swallowing, breathing by gills and drawing air into lungs.

Physiology The study of the function of the parts of living organisms.

Pituitary A small *endocrine gland* attached to the underside of the brain which secretes *gonadotrophins*, *prolactin*, *ACTH* and *oxytocin* as well as hormones that control body fluids and the thyroid gland.

Placenta An organ produced by the interaction between embryonic and maternal tissues. The placenta allows the embryo to gather nutrients from its mother and discard waste. The placenta is also an interface at which the two participants can communicate.

Polyspermy The abnormal entry of more than one sperm into an egg.

Preformation The theory that the entire body plan of an embryo is already established in either the sperm ('spermistic' preformation) or the egg ('ovistic' preformation) prior to fertilisation.

Progesterone A steroid produced by the *corpus luteum* and the *placenta*, if one is present. Progesterone participates in the menstrual cycle and is essential for pregnancy.

Prolactin A protein hormone secreted by the *pituitary* which is involved in the maintenance of *lactation*, as well as a diverse array of other processes in different species.

Prostaglandins A group of fat-derived chemical messengers produced by a variety of tissues. Among other things, prostaglandins are involved in the destruction of the *corpus luteum* and the process of labour.

Protein A molecule produced by the linking together of smaller units called amino acids. Proteins are extremely diverse and constitute most of the 'machinery' inside cells.

Pseudopregnancy (false pregnancy, phantom pregnancy) The occurrence of some or all of the physiological, psychological or anatomical changes of pregnancy in a non-pregnant female animal.

Recapitulation My lazy shorthand for the *biogenetic law*.

Receptors Molecules carried by many body cells, which bind to incoming messenger molecules (*hormones*, for example) and mediate the effects of those messengers on the cell.

Recessive gene A gene inherited from one parent which can be completely masked by the presence of a counterpart *dominant gene* inherited from the other parent. Thus it must be inherited from both parents if its effects are to become apparent.

Ribonucleic acid (RNA) A long molecule similar to *DNA* which is capable of carrying genetic information. In humans, however, RNA is mainly used to carry temporary copies of *genes* from the *nucleus* to the *cytoplasm*, and to make *proteins* from those copies.

RNA See *ribonucleic acid*.

Second-kidney Mesonephros.

Semen The sperm-containing fluid produced by male animals. Only a tiny fraction of mammalian semen is made up by sperm, the rest being contributed by the secretions of various male reproductive glands, such as the prostate.

Sensory plaques Ectodermal placodes.

Sex chromosomes Specialised *chromosomes* which carry the *genes* that determine the sex of an animal. In humans, possession of a Y chromosome usually leads to a male appearance. At least one X chromosome is essential for life, so 'normal' women are usually XX and men are XY.

Single-egg twins Also loosely called 'identical' or, more correctly, monozygotic, these twins are derived from a single egg.

Skin block Dermatome.

Somites Segmental blocks of *mesoderm* that form alongside the *neural tube*. The somites divide into portions that migrate around the body to form most bones, muscles and much of the skin.

Species Scientists like to classify animals into species, and a common definition of a species is a population of animals that can all breed to produce fertile offspring. There are problems with this definition, however, and in fact it has been extremely difficult to produce a universal definition of 'species'.

Spontaneous generation The idea that life can be created from non-living matter.

Steroids A diverse group of hormones made from cholesterol. They include *estrogen, androgens, progesterone, cortisol* and some other hormones that control blood pressure and body fluids.

Surfactant A detergent-like molecule produced by the lungs which reduces the amount of work the chest must do to expand the tiny air spaces in the lungs.

Taste plaque Epibranchial placode.

Thymic helper lymphocyte The master control-cell of the immune system that coordinates the immune responses of *antibody lymphocytes* and *thymic killer lymphocytes*, as well as many guard cells.

Thymic killer lymphocyte A cell of the immune system that kills cells when they have become infected by specific parasites, especially viruses.

Thymus Derived from *gill grooves*, and lying in front of the heart and lungs, the thymus is the organ in which *thymic helper* and *thymic killer lymphocytes* mature.

Trophoblast The outer part of the embryo at the *blastocyst* stage, which eventually forms the outer layer of the *placenta* and *fetal membranes*.

Umbilical cord The structure that connects the fetus' navel with the *placenta*. Probably derived from *mesoderm, ectoderm* and the *allantois*, it contains the large umbilical blood vessels.

Vein A blood vessel that carries blood towards the heart. Cf. *artery*.

Vertebrate An animal with a backbone. The modern–day vertebrates are the gristly and bony fish, amphibians, turtles, snakes and lizards, crocodiles, birds and mammals.

Virus An entity, usually much smaller than a *bacterium*, that is capable of reproducing only by infecting a bacterial or *eukaryotic* cell. The genes of a virus may be encoded by either *DNA* or *RNA*, and these are usually contained in a simple protein coat, sometimes within an additional fatty envelope.

Wolffian ducts The ducts that drain the second set of kidneys, but which can be adapted to form parts of the male reproductive system.

X chromosome See *sex chromosomes*.

Y chromosome See *sex chromosomes*.

Yolk sac A *fetal membrane*, actually a sac derived from the early *endoderm* which protrudes from the navel.

Zona pellucida The 'clear zone' is a glassy protein coat that surrounds the mammalian egg.

Zygote A cell formed by the fusion of sperm and egg.

Index

○ ○ ○ ○ ○ ○ ○ ○ ○

newborns
  cold sensitivity, 234–5
  first breath, 230
nipples, 236
nose, 140
notochord, 124, 128, 281
nucleus, 281
nutrition, *see* diet in pregnancy

odour preference test, 183
olfactory tract, 140
organ transplantation, 176–7, 178–81, 186
orgasm, 14
ovaries, 67, 69–70
  development, 149, 150–51, 158–9
ovulation, 76
  induced, 85, 88, 280
oxygen diffusion across placenta, 173–4
oxytocin, 90–91, 218, 223, 241, 282

Padua, 56
paedomorphosis, 111, 282
pain
  in childbirth, 215, 225
  fetal awareness of, 142
pancreas, 137
parasites, 32
parasitic twins, 122–3
parathyroid gland, 148
parthenogenesis, 20, 33–4, 35–6, 282
  in humans, 36–7
paternal genes, 36, 189–90
pea plants, 22–3, 254–5
pelvis, 223
penis, 155
perinatal death, 205–12
phantom pregnancy, 83, 84, 282
pharynx, 145, 146, 282
phototherapy, 197
physiology, 282
Pikaia, 127–8
pituitary gland, 70–71, 219, 282
placenta, 166–7, 168–74, 193–4, 282
  artificial, 252
  burying, 174
  deciduate, 171, 261

eating, 235–6
and initiation of birth, 221–2
Listeria infestation, 208
MHC, 190–92
multiple, 170
nutrient transport across, 172–3
oxygen diffusion across, 173–4
and paternal genes, 36
survival of, 263
tumours, 96
plant chloroplasts, 44
Pliny, 77
polar body, 16
pollution
  and endometriosis, 80
  by estrogen, 257
Polo, Marco, 216
polydactyly, 133–4
polyspermy, 15–16, 282
postnatal depression, 243–4
postnatal mood disorders, 242–3
postnatal psychosis, 244
postpartum blues, 243
poverty, 206
pre-eclampsia, 1–4
  susceptibility to, 4
  treatment, 3
preformation, 21–2, 46–9, 282
pregnancy tests, 94
premature babies, 229–30
premature labour, 216
Privithi, 17
progenitor cells, 200
progesterone, 74–5, 76, 81–2, 257–8, 282
  contraceptive use, 258
prolactin, 216, 239, 240, 282
prostaglandin, 90, 93, 218, 222, 282
protein, 282
pseudopregnancy, 83, 84, 282

radioimmunoassay, 69
rat
  artificial incubation, 252
  placenta, 263
  reproduction, 88
receptors, 282
recessive genes, 24, 25, 282–3